Student Solutions Manual for
Introductory Statistics: A Problem Solving Approach

Stephen Kokoska
Bloomsburg University

Julie Clark
Hollins College

©2011 by W.H. Freeman and Company

ISBN-13: 978-1-4292-4281-3
ISBN-10: 1-4292-4281-7

Printed in the United States of America

First printing

W.H. Freeman and Company
41 Madison Avenue
New York, NY 10010
Houndmills, Basingstoke
RG21 6XS England

www.whfreeman.com

Contents

Chapter 1: An Introduction to Statistics and Statistical Inference   1

Chapter 2: Tables and Graphs for Summarizing Data   2

Chapter 3: Numerical Summary Measures   13

Chapter 4: Probability   21

Chapter 5: Random Variables and Discrete Probability Distributions   31

Chapter 6: Continuous Probability Distributions   38

Chapter 7: Sampling Distributions   47

Chapter 8: Confidence Intervals Based on a Single Sample   54

Chapter 9: Hypothesis Tests Based on a Single Sample   65

Chapter 10: Confidence Intervals and Hypothesis Tests Based on Two Samples or Treatments   75

Chapter 11: The Analysis of Variance   84

Chapter 12: Correlation and Linear Regression   91

    Optional Sections:

    Section 12.6: The Polynomial and Qualitative Predictor Models   107

    Section 12.7: Model Selection Procedures   110

Chapter 13: Categorical Data and Frequency Tables   115

Chapter 14: Nonparametric Statistics   121

Chapter 1: An Introduction to Statistics and Statistical Inference

Section 1.2: Populations, Samples, Probability, and Statistics

1.1 a. Descriptive. **b.** Inferential. **c.** Inferential. **d.** Inferential. **e.** Descriptive. **f.** Descriptive.

1.3 a. Open-heart patients operated on in the last year. **b.** 30 patients selected. **c.** Length of stay.

1.5 Population: employees at Citigroup Inc. Sample: 35 employees selected.

1.7 Population: 10,000 families affected by the flood. Sample: 75 affected families selected.

1.9 a. Population: All Americans. **b.** Sample: 1000 Americans selected. **c.** Variable: Whether or not each believes sharks are dangerous.

1.11 Population: People diagnosed with hepatitis C. Sample: 50 patients selected. Variable: Liver enzyme levels.

1.13 a. Population: all cheddar cheeses. Sample: 20 cheddar cheeses selected. **b.** Answers will vary. Probability question: What is the probability at least ten of the cheddar cheeses selected are aged less than two years? Statistics questions: Suppose 12 of the cheddar cheeses selected are aged less than two years. Does this suggest that the true proportion of all cheddars aged less than two years has decreased?

1.15 a. Population: All American companies. **b.** Sample: 75 companies selected. **c.** Variable: Whether each company has overseas IT workers. **d.** Answers will vary. Probability question: What is the probability exactly 30 of the 75 companies selected have overseas IT workers? Statistics question: Use the resulting data to determine if there is evidence the proportion of companies with overseas IT workers has changed.

Section 1.3: Experiments and Random Samples

1.17 a. Observational study. **b.** The sample consists of the 25 volunteer fire companies selected. **c.** This is not a random sample. Only the largest companies were selected.

1.19 Assign a number to each shipped weather station. Select numbers using a random number generator and examine each weather station corresponding to the numbers selected.

1.21 a. Population: All men who use a disposable razor. Sample: 100 men selected. **b.** This is not a random sample. Only men observed buying a razor were selected.

1.23 Assign a number to each challenge. Randomly select numbers from a random number table or random number generator and consider the challenge corresponding to each number.

1.25 a. Experimental study. **b.** Variable: Lifetime of each blossom. **c.** Flip a coin: heads is treated, tails is untreated.

1.27 a. Population: All ceramic tile from this manufacturer. Sample: 25 tiles selected. **b.** Not a random sample, all tiles from the same box.

Chapter Exercises

1.29 a. Descriptive. **b.** Descriptive. **c.** Inferential. **d.** Inferential.

1.31 Population: All teenagers. Sample: Teenagers obtained. Variable: Whether each teenager can cook.

1.33 Population: All individuals. Sample: the 78 people selected. Variable: cortisol level.

1.35 a. Observational study. **b.** The sample consists of the people who rode the train and were asked if they used the Sightseer Lounge. **c.** This is not a random sample. Assign a number to each passenger. Select numbers using a random number generator and ask each passenger corresponding to the numbers selected.

1.37 a. Every installed software title. **b.** 1000 software titles selected. **c.** Whether the software title is pirated. **d.** Answers will vary. Probability question: What is the probability more than 250 of the 1000 software titles selected are pirated? Statistics question: Use the resulting data to determine whether the software piracy rate in the United States has decreased.

Chapter 2: Tables and Graphs for Summarizing Data

Section 2.1: Types of Data

2.1 a. Numerical, continuous. **b.** Numerical, discrete. **c.** Categorical. **d.** Numerical, discrete. **e.** Numerical, continuous. **f.** Categorical.

2.3 a. Numerical, discrete. **b.** Numerical, discrete. **c.** Categorical. **d.** Numerical, continuous. **e.** Numerical, continuous. **f.** Categorical.

2.5 a. Continuous. **b.** Continuous. **c.** Discrete. **d.** Continuous. **e.** Continuous. **f.** Discrete.

2.7 a. Continuous. **b.** Discrete. **c.** Discrete. **d.** Continuous. **e.** Discrete. **f.** Discrete.

2.9 a. Discrete. **b.** Categorical. **c.** Continuous. **d.** Continuous. **e.** Categorical. **f.** Continuous.

Section 2.2: Bar Charts and Pie Charts

2.11 Frequency distribution:

Art	Frequency	Relative frequency
Abstract	15	0.3571
Expressionist	6	0.1429
Realist	12	0.2857
Surrealist	9	0.2143
Total	42	1.0000

2.13 a. Frequency distribution:

County	Frequency	Relative frequency
Adair	915	0.0946
Carroll	1081	0.1118
Chariton	1095	0.1132
Grundy	735	0.0760
Linn	969	0.1002
Livingston	903	0.0934
Macon	1351	0.1397
Mercer	569	0.0588
Putnam	723	0.0748
Schuyler	480	0.0496
Sullivan	850	0.0879
Total	9671	1.0000

b. Bar chart:

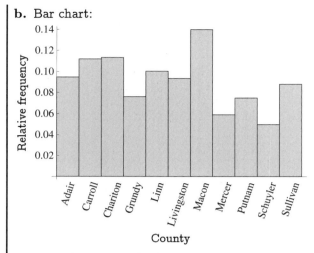

2.15 a. Frequency distribution:

Political affiliation	Frequency	Relative frequency
Democrat	23	0.3833
Republican	18	0.3000
Independent	19	0.3167
Total	60	1.0000

b. Pie chart:

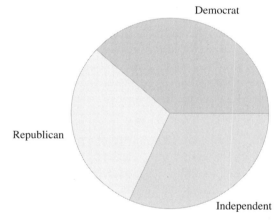

2.17 a. Frequency distribution:

Grade	Frequency	Relative frequency
A	10	0.06757
B	43	0.29054
C	54	0.36486
D	26	0.17568
F	15	0.10135
Total	148	1.00000

b. Bar chart:

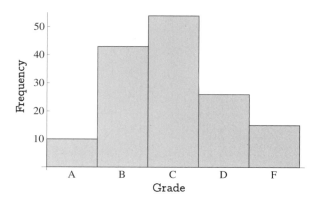

Pie chart:

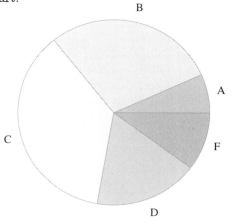

c. There were 148 students in this psychology class.
Proportion of students who passed
$$= 0.06757 + 0.29054 + 0.36486 + 0.17568$$
$$= 1 - 0.10135 = 0.89765$$

2.19 a. Frequency distribution:

Table saw brand	Frequency	Relative frequency
Black & Decker	4	0.1429
Craftsman	5	0.1786
Delta	6	0.2143
DeWalt	7	0.2500
Makita	6	0.2143
Total	28	1.0000

b. Bar chart:

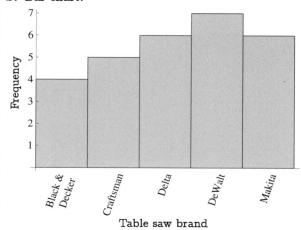

Pie chart:

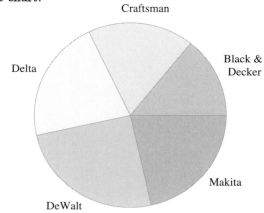

c. Proportion who use Craftsman or Black &
Decker $= 0.1429 + 0.1786 = 0.3215$ **d.** Proportion
who do not use Delta $= 1 - 0.2143 = 0.7857$

2.21 a. Frequency distribution:

Book type	Frequency	Relative frequency
Education	5	0.1667
Law	3	0.1000
Literature	4	0.1333
Medicine	7	0.2333
Science	5	0.1667
Technology	6	0.2000

4

Chapter 2. Tables and Graphs for Summarizing Data

b. Bar chart:

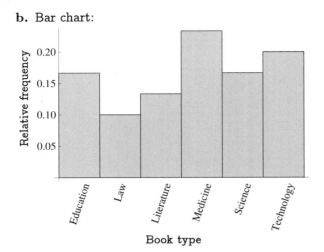

Pie chart:

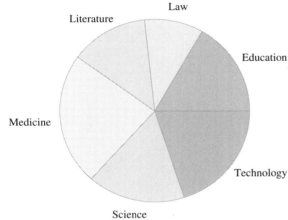

c. The classification with the largest proportion of books borrowed was medicine. Since these books appear to be in demand, the public library should consider purchasing more books in this subject area. Note that no type of book was overwhelming borrowed.

2.23 a. Bar chart:

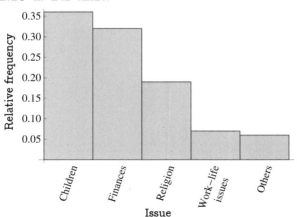

Pie chart:

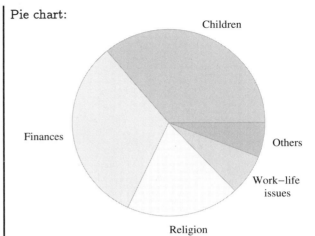

b. Frequency distribution:

Issue		Frequency	Relative frequency
Children	$(1002)(0.36) =$	361	0.36
Finances	$(1002)(0.32) =$	321	0.32
Religion	$(1002)(0.19) =$	190	0.19
Work-life issues	$(1002)(0.07) =$	70	0.07
Others	$(1002)(0.06) =$	60	0.06

2.25 a. Frequency distribution:

Think tank	Frequency	Relative frequency
Brookings Institution	2380	0.3167
Council on Foreign Relations	1191	0.1585
Heritage Foundation	1168	0.1554
RAND Corporation	740	0.0985
Cato Institute	640	0.0852
Urban Institute	558	0.0743
Carter Center	341	0.0454
Carnegie Endowment	287	0.0382
Aspen Institute	209	0.0278
Total	7514	1.0000

b. Bar chart:

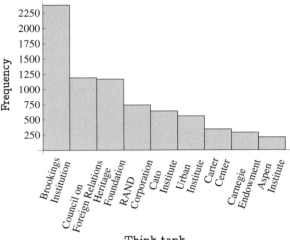

Think tank

Pie chart:

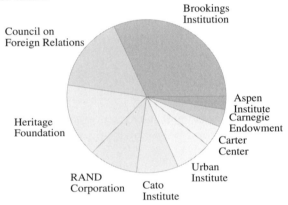

2.27 Frequency distribution:

Casino		Frequency		Relative frequency
Bally's		40	40/200 =	0.200
Caesars		25		0.125
Harrah's		32	32/200 =	0.160
Resorts	(200)(0.110) =	22		0.110
Sands		25	25/200 =	0.125
Trump Plaza	(200)(0.28) =	56		0.280
Total		200		1.000

a. $\frac{25}{n} = 0.125 \implies n = \frac{25}{0.125} = 200$

b. The Trump Plaza is the most preferred casino by people in this survey. This class had the largest frequency and relative frequency.

2.29 a. Frequency distribution:

Response		Frequency		Relative frequency
Excellent		50	50/1000 =	0.0500
Very good		152	152/1000 =	0.1520
Good		255	255/1000 =	0.2550
Fair	(1000)(0.4250) =	425		0.4250
Poor	(1000)(0.1180) =	118		0.1180
Total		1000		1.0000

b. Bar chart:

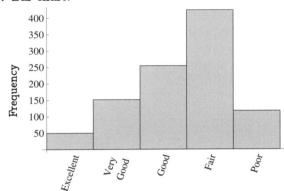

Response

Pie chart:

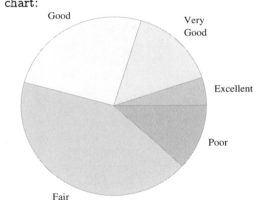

c. $1 - (0.2550 + 0.0500) = 1 - 0.2020 = 0.7980$

2.31 a. Frequency distribution:

Response	Frequency	Relative frequency
Misc. Recreation	4	0.0130
In bounds skier	4	0.0130
Snowshoer	20	0.0649
Snowboarder	34	0.1104
Out of bounds skier	30	0.0974
Backcountry skier	49	0.1591
Snowmobiler	126	0.4091
Climber/hiker	32	0.1039
Highway personnel	1	0.0032
Others at work	2	0.0065
Resident	6	0.0195
Total	308	1.0000

b. Bar chart using frequency:

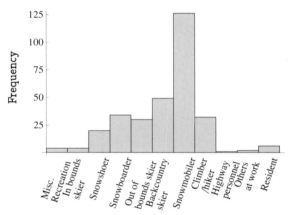

Bar chart using relative frequency:

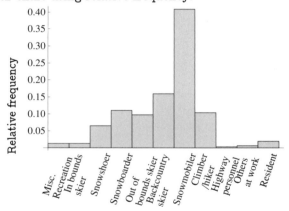

The two graphs are identical, except for the label on the vertical axis. Relative frequency might be more useful for avalanche rescuers.

2.33 a. Frequency distribution:

	Men		Women	
Rating	Frequency	Relative frequency	Frequency	Relative frequency
Excellent	368	0.1840	350	0.2333
Very good	550	0.2750	375	0.2500
Good	426	0.2130	165	0.1100
Fair	450	0.2250	360	0.2400
Poor	206	0.1030	250	0.1667
Total	2000	1.0000	1500	1.0000

b. Side-by-Side bar chart:

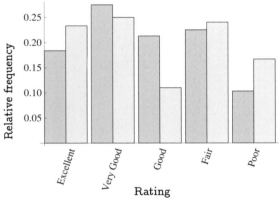

c. The total number of men who responded is different from the total number of women who responded.

Section 2.3: Stem-and-Leaf Plots

2.35 Stem-and-leaf plot:

```
2 | 79
3 |
3 | 56669
4 | 112
4 | 779
5 | 01124
5 | 57789
6 | 1444
6 | 68        Stem = 1
7 | 1         Leaf = 0.1
```

The center of the data is between 5.0 and 5.5. A typical value is 5.1.

2.37 Stem-and-leaf plot:

53	0344
53	799
54	111344
54	566677777889
55	112334
55	67777
56	002
56	9

Stem = 10
Leaf = 1

The center of the data is between 545 and 550. A typical value is 547.

2.39 a. 543, 543, 549. **b.** 574 **c.** The data tail off slowly on the low end. **d.** There do not appear to be any outliers.

2.41 a. Stem-and-leaf plot, truncated leaves:

0	4
1	16
2	12779
3	3449
4	0111125555799
5	699
6	023444
7	0123679
8	258
9	3

Stem = 1
Leaf = 0.1

b. Stem-and-leaf plot, rounded leaves:

0	4
1	16
2	2378
3	0355
4	001222256668
5	00799
6	133445
7	1134779
8	269
9	3

Stem = 1
Leaf = 0.1

c. These two graphs are slightly different. However, they both suggest the same general shape, center, and variability. A typical value is 4.7 or 4.8.

2.43 a. Back-to-back stem-and-leaf plot:

Lower floors		Upper floors
	10	14
	10	
4300	11	1
99887777777666665555	11	55578
33333222211111110000000	12	1244
665	12	55556777899
0	13	122234
	13	567888999
	14	02234
	14	5888
	15	244

Stem = 10
Leaf = 1

b. The lower floors distribution is more compact and has, on average, smaller values. The upper floors distribution has more variability and has, on average, larger values.

2.45 a. Stem-and-leaf plot:

1	00699
2	244557
3	11245678899
4	000011555669
5	2259
6	8
7	1

Stem = 10
Leaf = 1

b. A typical value is approximately 38.5. There are no outliers.

2.47 a. Stem-and-leaf plot:

2	135667
3	13456777788999
4	0001122355
5	002245566
6	
7	
8	0

Stem = 10
Leaf = 1

b. A typical Texaco station has been in operation for 40 years. There is one outlier: 80.

2.49 a. Stem-and-leaf plot:

0	899
1	0
1	333
1	455
1	67777
1	8
2	001
2	233
2	45
2	
2	
3	
3	3

Stem = 10
Leaf = 1

b. A typical weight is approximately 17. There is one outlier: 33.

2.51 a. Back-to-back stem-and-leaf plot:

With		Without
	250	24
	251	
84	252	145
3	253	788
63	254	49
8742221	255	023
9866653110	256	149
873	257	1248
6553	258	23367
4	259	3
	260	26
	261	2
	262	3

Stem = 1
Leaf = 0.1

b. The "With" distribution is unimodal, compact, and approximately symmetric. The "Without" distribution is unimodal, has lots of variability, and is slightly negatively skewed. It appears the humidifier does help a piano stay in tune. The "With" humidifier distribution is more compact and centered near 256.

Section 2.4: Frequency Distributions and Histograms

2.53 Frequency distribution:

Class	Frequency	Relative frequency	Cumulative relative frequency
78–80	2	0.050	0.050
80–82	4	0.100	0.150
82–84	4	0.100	0.250
84–86	4	0.100	0.350
86–88	9	0.225	0.575
88–90	6	0.150	0.725
90–92	9	0.225	0.950
92–94	2	0.050	1.000
Total	40	1.000	

2.55 Frequency histogram:

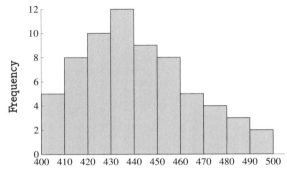

2.57 Frequency distribution:

Class	Frequency	Relative frequency	Cumulative relative frequency
100–150	155	0.1938	0.1938
150–200	120	0.1500	0.3438
200–250	130	0.1625	0.5063
250–300	145	0.1813	0.6875
300–350	150	0.1875	0.8750
350–400	100	0.1250	1.0000
Total	800	1.0000	

2.59 Frequency distribution:

Class	Frequency	Relative frequency	Cumulative relative frequency
0–25	$1000 * 0.150 = 150$	0.150	0.150
25–50	$1000 * 0.200 = 200$	$0.350 - 0.150 = 0.200$	0.350
50–75	$1000 * 0.175 = 175$	$0.525 - 0.350 = 0.175$	0.525
75–100	$1000 * 0.150 = 150$	$0.675 - 0.525 = 0.150$	0.675
100–125	$1000 * 0.125 = 125$	$0.800 - 0.675 = 0.125$	0.800
125–150	$1000 * 0.100 = 100$	$0.900 - 0.800 = 0.100$	0.900
150–175	$1000 * 0.075 = 75$	$0.975 - 0.900 = 0.075$	0.975
175–200	$1000 * 0.025 = 25$	$1.000 - 0.975 = 0.025$	1.000
Total	1000	1.000	

Frequency histogram:

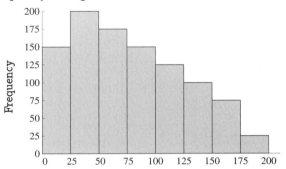

2.61 a. Frequency distribution:

Class	Frequency	Relative frequency	Cumulative relative frequency
0–10	1	0.0167	0.0167
10–20	3	0.0500	0.0667
20–30	9	0.1500	0.2167
30–40	11	0.1833	0.4000
40–50	12	0.2000	0.6000
50–60	10	0.1667	0.7667
60–70	7	0.1167	0.8833
70–80	5	0.0833	0.9667
80–90	2	0.0333	1.0000
Total	60	1.0000	

Frequency histogram:

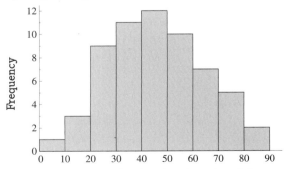

b. The shape of the distribution is unimodal,

bell-shaped, and symmetric. **c.** $M \approx 43$
d. $Q_1 \approx 31.5$ **e.** $Q_3 \approx 58.5$

2.63 a. Frequency distribution:

Class	Frequency	Relative frequency	Cumulative relative frequency
0–50	5	0.1667	0.1667
50–100	9	0.3000	0.4667
100–150	5	0.1667	0.6333
150–200	3	0.1000	0.7333
200–250	2	0.0667	0.8000
250–300	1	0.0333	0.8333
300–350	4	0.1333	0.9667
350–400	0	0.0000	0.9667
400–450	0	0.0000	0.9667
450–500	0	0.0000	0.9667
500–550	1	0.0333	1.0000
Total	30	1.0000	

Frequency histogram:

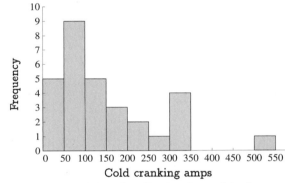

b. The shape of the distribution is positively skewed. **c.** $M \approx 110$

2.65 a. Relative frequency histogram, United States:

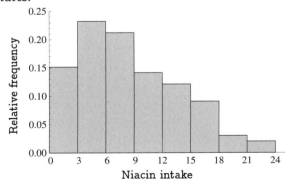

Relative frequency histogram, Europe:

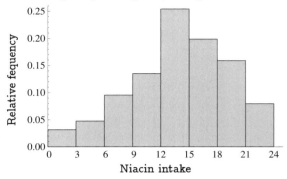

b. The shape of the United States distribution is unimodal and positively skewed. The shape of the Europe distribution is unimodal and negatively skewed. On average, it appears Europeans have a greater daily niacin intake.

2.67 a. Frequency histogram:

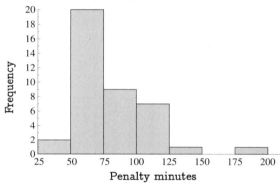

b. The shape of the distribution is slightly positively skewed. The center is approximately 73 and there is little variability. There is one outlier: 182. **c.** $m \approx 117.5$

2.69 a. Frequency distribution:

Class	Frequency	Relative frequency	Cumulative relative frequency
100–105	10	0.050	0.050
105–110	75	0.375	0.425
110–115	40	0.200	0.625
115–120	25	0.125	0.750
120–125	20	0.100	0.850
125–130	15	0.075	0.925
130–135	10	0.050	0.975
135–140	5	0.025	1.000
Total	200	1.000	

b. Frequency histogram:

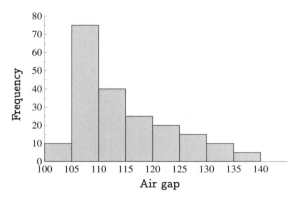

c. Proportion of air gaps between 110 and 125 $= 0.200 + 0.125 + 0.100 = 0.850 - 0.425 = 0.425$

2.71 a. Frequency histogram:

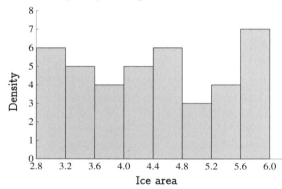

b. There is no clear shape to the distribution. The center is approximately 4.35 and there appears to be a lot of variability. **c.** $Q_1 \approx 3.45$, $Q_3 \approx 5.25$.
d. There should be $0.5 * 40 = 20$ values between Q_1 and Q_3. There are 20 values between Q_1 and Q_3.

Chapter Exercises

2.73 a. Stem-and-leaf plot:

```
1 | 5789
2 | 01134
2 | 5567889999
3 | 0011223344
3 | 55566679
4 | 01234
4 | 89          Stem = 0.1
5 | 2           Leaf = 0.01
```

b. The shape of the distribution is unimodal and slightly skewed to the right. The center is approximately 0.31, and there is little variability. There is one possible outlier: 0.52.

2.75 a. Frequency distribution:

Class	Frequency	Relative frequency	Cumulative relative frequency
65–70	1	0.02	0.02
70–75	4	0.08	0.10
75–80	7	0.14	0.24
80–85	10	0.20	0.44
85–90	12	0.24	0.68
90–95	15	0.30	0.98
95–100	1	0.02	1.00
Total	50	1.0000	

b. Frequency histogram:

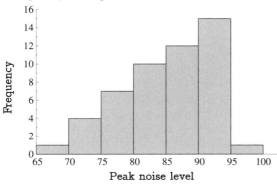

c. $0.02 + 0.08 + 0.14 = 0.24$

d. $0.30 + 0.02 = 0.32$

2.77 a. Back-to-back stem-and-leaf plot:

New		Traditional
	0	9
	1	5
87660	2	6
765443200	3	34
99774444322	4	2568
110	5	57
43	6	288
	7	3357
	8	358
	9	18
	10	03
	11	
	12	033
	13	3
	14	16

Stem = 1
Leaf = 0.1

b. The new equipment times tend to be smaller, the distribution is more compact. The traditional equipment times are more spread out and tend to be larger. **c.** The new equipment times tend to be better, shorter response times. The majority of the times are less than the traditional equipment times.

2.79 a. Frequency distribution:

Class	Frequency	Relative frequency	Cumulative relative frequency
000–100	3303	0.3820	0.3820
100–200	3613	$0.7999 - 0.3820 = 0.4179$	0.7999
200–300	1117	$0.9291 - 0.7999 = 0.1292$	0.9291
300–400	375	$0.9725 - 0.9291 = 0.0434$	0.9725
400–500	128	$0.9873 - 0.9725 = 0.0148$	0.9873
500–600	61	$0.9944 - 0.9873 = 0.0071$	0.9944
600–700	29	$0.9978 - 0.9944 = 0.0034$	0.9978
700–800	10	$0.9990 - 0.9978 = 0.0012$	0.9990
800–900	9	$1.0000 - 0.9990 = 0.0010$	1.0000
Total	8646		

To find each class frequency, multiply 8646 by the corresponding class relative frequency.
Frequency histogram:

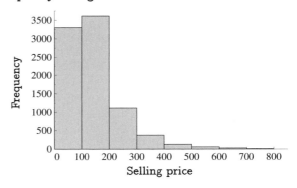

b. The shape of the distribution is positively skewed. The center is approximately 150 and there is a lot of variability. **c.** A typical selling price is 150 thousand dollars. There are a few possible outliers in the 800–900 class.

d. $1 - 0.9291 = 0.0709$

2.81 a. Bar chart, placebo:

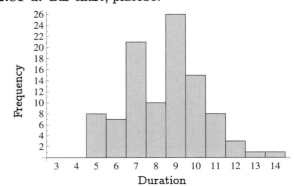

Bar chart: vitamin C:

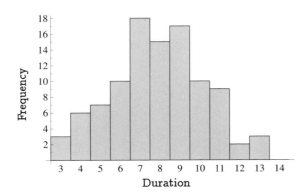

b. Both graphs appear to be centered at about the same duration. Both appear to be symmetric and bell-shaped. The placebo durations are slightly more compact than the vitamin C durations.
c. There is no graphical evidence to suggest vitamin C reduced the duration.

2.83 a. Frequency distribution:

Class	Frequency	Relative frequency	Cumulative relative frequency
100–110	1	0.02	0.02
110–120	2	0.04	0.06
120–130	9	0.18	0.24
130–140	7	0.14	0.38
140–150	9	0.18	0.56
150–160	9	0.18	0.74
160–170	11	0.22	0.96
170–180	1	0.02	0.98
180–190	1	0.02	1.00
Total	50	1.00	

Ogive:

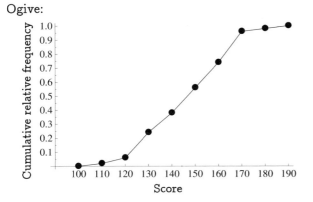

Chapter 3: Numerical Summary Measures

Section 3.1: Measures of Central Tendency

3.1 a. $\sum x_i = -15 + 6 + 40 + 13 + 38 = 82$

b. $\sum x_i^2 = (-15)^2 + 6^2 + 40^2 + 13^2 + 38^2$
$$= 225 + 36 + 1600 + 169 + 1444 = 3474$$

c. $\sum (x_i - 10) = -25 + -4 + 30 + 3 + 28 = 32$

d. $\sum (x_i - 5)^2 = (-20)^2 + 1^2 + 35^2 + 8^2 + 33^2$
$$= 400 + 1 + 1225 + 64 + 1089$$
$$= 2779$$

e. $\sum (2x_i) = -30 + 12 + 80 + 26 + 76 = 164$

f. $2 \sum x_i = 2(-15 + 6 + 40 + 13 + 38)$
$$= 2(82) = 164$$

3.3 a. $\bar{x} = 1057/10 = 105.7$

b. $\bar{x} = 356/27 = 13.1852$ **c.** $\bar{x} = 250.5/36 = 6.9583$

d. $\bar{x} = 1.355/11 = 0.1232$

e. $\bar{x} = -37.4/15 = -2.4933$

f. $\bar{x} = 496.81/28 = 17.7432$

3.5 a. $\bar{x} = (5 + 3 + \cdots + 7)/9 = 6.6667$; $n = 9$;
median is in position 5 in the ordered list, $\tilde{x} = 7$.

b. $\bar{x} = (-7 + 10 + \cdots + (-11))/11 = 6.6364$;
$n = 11$; median is in position 6, $\tilde{x} = 9$.

c. $\bar{x} = (5.4 + 3.3 + \cdots + 4.6)/9 = 10.6889$; $n = 9$;
median is in position 5, $\tilde{x} = 7.7$.

d. $\bar{x} = ((-103.7) + \cdots + (-115.6))/5 = -107.69$;
$n = 5$; median is in position 3, $\tilde{x} = -109.1$.

3.7 a. $\bar{x} < \tilde{x}$; suggests the distribution is skewed
left.   **b.** $\bar{x} \approx \tilde{x}$; suggests the distribution is
approximately symmetric.   **c.** $\bar{x} < \tilde{x}$; suggests the
distribution is skewed left.   **d.** $\bar{x} < \tilde{x}$; suggests the
distribution is skewed left.

3.9 a. $M = 6$ is the number that occurs most
often.   **b.** $M = 0$ is the number that occurs most
often.   **c.** There is no mode. Each observation
occurs exactly once.

3.11 a. $\bar{x} = \frac{1}{21}(81 + \cdots + 61) = 68.5238$

b. $n = 21$, the sample median is in position 11 of
the ordered list. $\tilde{x} = 67$.   **c.** Since $\tilde{x} < \bar{x}$, the
distribution is slightly skewed to the right.

3.13 a. $\bar{x} = \frac{1}{18}(6.5 + \cdots + 6.57) = 6.5083$
$n = 18$, the median is in position 9.5 in the ordered
list. $\tilde{x} = \frac{1}{2}(6.5 + 6.53) = 6.5150$

b. $\bar{y} = \frac{1}{18}(6.5 + \cdots + 10.57) = 6.7306$
The median is still in position 9.5 in the ordered
list. $\tilde{y} = \frac{1}{2}(6.5 + 6.53) = 6.5150$
The mean is pulled in the direction of the outlier
(10.57). The median remains the same.

3.15 a. $\bar{x} = \frac{1}{40}(700 + \cdots + 730) = 619.5$
$n = 40$, the median is in position 20.5 in the ordered
list. $\tilde{x} = \frac{1}{2}(620 + 620) = 620$.   **b.** $p = 0.05$;
$np = (40)(0.05) = 2$. Trim 2 observations from each
end of the ordered list; trim 500, 500, 740, 750.
$\bar{x}_{\text{tr}(0.05)} = \frac{1}{36}(500 + \cdots + 730) = 619.1667$

3.17 a. $\bar{x} = \frac{1}{20}(5729 + \cdots + 1120) = 100539.5$
$n = 20$, the median is in position 10.5 in the ordered
list. $\tilde{x} = \frac{1}{2}(9673 + 10598) = 10135.5$ **b.** The
median is a better measure of central tendency since
there are extreme outliers.

3.19 a. $\bar{x} = \frac{1}{30}(54 + \cdots + 72) = 59.8667$
$n = 30$, the median is in position 15.5 in the ordered
list. $\tilde{x} = \frac{1}{2}(60 + 61) = 60.5$ **b.** $p = 0.10$;
$np = (30)(0.10) = 3$. Trim 3 observations from each
end of the ordered list; trim 46, 46, 50, 68, 70, 72.
$\bar{x}_{\text{tr}(0.10)} = \frac{1}{24}(51 + \cdots + 68) = 60.1667$ **c.** There are
two modes: 60 and 61. Both of these observations
occur 4 times.

3.21 a. $\bar{x} = \frac{1}{46}(7.5 + \cdots + 6.0) = 6.4565$
$n = 46$, the median is in position 23.5 in the ordered
list. $\tilde{x} = \frac{1}{2}(6.5 + 6.5) = 6.5$ **b.** $M = 6.5$
c. $\bar{y} = \frac{1}{46}(20.25 + \cdots + 16.20) = 17.4326 = 2.7\bar{x}$

3.23 a. $\bar{x} = \frac{1}{26}(4059.2 + \cdots + 9.4) = 1480.2692$
b. $\bar{y} = \frac{1}{26}(48710.4 + \cdots + 112.8) = 17763.2308 = 12\bar{x}$

3.25 a. $\bar{x} = \frac{1}{17}(8.9 + \cdots + 12.4) = 16.5588$
$n = 17$, the median is in position 9 in the ordered
list. $\tilde{x} = 15.1$ **b.** Since $\tilde{x} < \bar{x}$, the distribution is
slightly skewed to the right.   **c.** If the maximum
value is changed to a bigger number, the median
remains the same, and the mean increases. If the
maximum value is changed to a smaller number, the
median remains the same, until the new number is
smaller than 15.1. The smallest the new value can
be is 0. The mean will always be larger than 13.6.
Therefore, the maximum value cannot be changed
to a new value so that $\bar{x} = \tilde{x}$.

3.27 a. $\bar{x} = 6883.4 = \frac{1}{5}(1078 + \cdots + x_5)$
$6883.4 * 5 = 34417.0 = 1078 + 5833 + 10772 + 7320 + x_5$
$34417.0 - 1078 - 5833 - 10772 - 7320 = 9414 = x_5$

b. The sample mean for the first four observations
is 6250.75. Let $x_5 = 6250.75$. The sample mean
remains the same, and this would be the sample
median since 6250.75 is the third number in the
ordered list.

3.29 a. $\bar{x} = \frac{1}{24}(58 + \cdots + 63) = 57.1667$

$n = 24$, the median is in position 12.5 in the ordered list. $\tilde{x} = \frac{1}{2}(57 + 58) = 57.5$

b. $\bar{y} = \frac{1}{24}(14.4444 + \cdots + 17.2222) = 13.9815$

c. $\bar{y} = (\bar{x} - 32)/1.8$

Section 3.2: Measures of Variability

3.31 a. $R = 6.6 - 2.7 = 3.9$;
$s^2 = \frac{1}{9}(252.81 - \frac{1}{10}(2332.89)) = 2.1690$;
$s = \sqrt{2.1690} = 1.4728$

b. $R = 36.3 - 15.7 = 20.6$;
$s^2 = \frac{1}{14}(7345.82 - \frac{1}{15}(104458.24)) = 27.2812$;
$s = \sqrt{27.2812} = 5.2231$

c. $R = 37.09 - (-61.23) = 98.32$;
$s^2 = \frac{1}{7}(7710.9316 - \frac{1}{8}(296.5284)) = 1096.2665$;
$s = \sqrt{1096.2665} = 33.1099$

d. $R = 3.13 - (-2.57) = 5.7$;
$s^2 = \frac{1}{19}(50.2207 - \frac{1}{20}(22.3729)) = 2.5843$;
$s = \sqrt{2.5843} = 1.6076$

3.33 a. Q_1: $d_1 = 60/4 = 15 \rightarrow$ depth $= 15.5$
Q_3: $d_3 = 3(60)/4 = 45 \rightarrow$ depth $= 45.5$

b. Q_1: $d_1 = 37/4 = 9.25 \rightarrow$ depth $= 10$
Q_3: $d_3 = 3(37)/4 = 27.75 \rightarrow$ depth $= 28$

c. Q_1: $d_1 = 100/4 = 25 \rightarrow$ depth $= 25.5$
Q_3: $d_3 = 3(100)/4 = 75 \rightarrow$ depth $= 75.5$

d. Q_1: $d_1 = 48/4 = 12 \rightarrow$ depth $= 12.5$
Q_3: $d_2 = 3(48)/4 = 36 \rightarrow$ depth $= 36.5$

3.35 a. $s^2 = \frac{1}{9}(16690 - \frac{1}{10}(128164)) = 430.4$
$s = \sqrt{430.4} = 20.7461$

b. $s^2 = \frac{1}{9}(6370 - \frac{1}{10}(24964)) = 430.4$
$s = \sqrt{430.4} = 20.7461$

The sample variance and the sample standard deviation are the same in parts (a) and (b).

c. $s^2 = \frac{1}{9}(6676000 - \frac{1}{10}(51265600)) = 172160$
$s = \sqrt{172160} = 414.9217$

The sample variance is multiplied by $20^2 = 400$, and the sample standard deviation is multiplied by 20.

3.37 a. $R = 31.16 - 29.62 = 1.54$

b. $s^2 = \frac{1}{9}(9223.8527 - \frac{1}{10}(92215.4689)) = 0.2562$
$s = \sqrt{0.2562} = 0.5062$

c. Q_1; depth $= 3$; $Q_1 = 29.98$
Q_3; depth $= 8$; $Q_3 = 30.8$
$IQR = 30.8 - 29.98 = 0.82$

3.39

a. $s^2 = \frac{1}{15}(7580975 - \frac{1}{16}(121110025)) = 773.2292$
$s = \sqrt{773.2292} = 27.8070$

b. Q_1; depth $= 4.5$; $Q_1 = \frac{1}{2}(665 + 670) = 667.5$
Q_3; depth $= 12.5$; $Q_3 = \frac{1}{2}(700 + 700) = 700$

c. $IQR = 700 - 667.5 = 32.5$
$QD = (700 - 667.5)/2 = 16.25$

3.41 a. $\bar{x} = \frac{1}{7}(51 + \cdots + 79) = 62.1429$
$s^2 = \frac{1}{6}[(51 - 62.1429)^2 + \cdots + (79 - 62.1429)^2]$
$= \frac{1}{6}(948.8571) = 158.1429$

b. $s^2 = \frac{1}{6}(27981 - \frac{1}{7}(189225)) = 158.1429$

c. The answers in parts (a) and (b) are the same.

3.43 a. Q_1; depth $= 3$; $Q_1 = 291$
Q_3; depth $= 8$; $Q_3 = 313$
$IQR = 313 - 291 = 22$

b. $s^2 = \frac{1}{9}(944864 - \frac{1}{10}(9400356)) = 536.4889$
$s = \sqrt{536.4889} = 23.1622$

c. Q_1; depth $= 3$; $Q_1 = 291$
Q_3; depth $= 8$; $Q_3 = 313$
$IQR = 313 - 291 = 22$
$s^2 = \frac{1}{9}(915851 - \frac{1}{10}(9054081)) = 1160.3222$

d. IQR is the same, s^2 is larger. s^2 is more sensitive to outliers.

3.45 a. $s^2 = \frac{1}{18}(1732729 - \frac{1}{19}(28826161))$
$= 11975.7018$
$s = \sqrt{11975.7018} = 109.4335$

b. Q_1; depth $= 5$; $Q_1 = 265$
Q_3; depth $= 15$; $Q_3 = 352$
$IQR = 352 - 265 = 87$

c. $s^2 = \frac{1}{16}(1732172 - \frac{1}{17}(28472896)) = 3580.9853$
$s = \sqrt{3580.9853} = 59.8413$
Q_1; depth $= 5$; $Q_1 = 270$
Q_3; depth $= 13$; $Q_3 = 352$
$IQR = 352 - 270 = 82$

Both values are smaller in the modified data set. By eliminating the two smallest values (outliers), the variability in the modified data set is smaller.

3.47 a. Q_1; depth $= 8.5$;
$Q_1 = \frac{1}{2}(1.12 + 1.13) = 1.125$
Q_3; depth $= 24.5$; $Q_3 = \frac{1}{2}(2 + 2.01) = 2.005$
$IQR = 2.005 - 1.125 = 0.88$

b. 1.12. If the minimum stream velocity is changed to a value greater than 1.12, then the number in the 8th position in the ordered list would change. This would change the value of Q_1.

c. $CQV = 100\frac{2.005-1.125}{2.005+1.125} = 28.115$

3.49 a. Q_1; depth = 3; $Q_1 = 300$

Q_3: depth = 8; $Q_3 = 401$

$IQR = 401 - 300 = 101$

b. Q_1; depth = 3; $Q_1 = 300$

Q_3: depth = 8; $Q_3 = 401$

$IQR = 401 - 300 = 101$

c. The maximum price could be lowered to 401. Anything lower would change the value in the 8th position in the ordered list, and therefore, IQR.

d. The minimum price could be raised to 300. Anything higher would change the value in the 3rd position in the ordered list, and therefore, IQR.

3.51 a. $\bar{x} = \frac{1}{6}(25 + \cdots + 45) = 29.6667$

Deviations about the mean:

$-4.6667, 9.3333, -13.6667, 5.3333, -11.6667, 15.3333$

b. $\sum(x_i - \bar{x}) = -4.6667 + \cdots + 15.3333 = 0$

c. $\sum(x_i - \bar{x}) = \sum x_i - \sum \bar{x}$

$\qquad = \sum x_i - n\bar{x}$

$\qquad = \sum x_i - n\frac{1}{n}\sum x_i$

$\qquad = \sum x_i - \sum x_i = 0$

3.53

$$\frac{1}{n-1}\sum(x_i - \bar{x})^2 = \frac{1}{n-1}\sum(x_i^2 - 2x_i\bar{x} + \bar{x}^2)$$
$$= \frac{1}{n-1}\left[\sum x_i^2 - 2\bar{x}\sum x_i + n\bar{x}^2\right]$$
$$= \frac{1}{n-1}\left[\sum x_i^2 - 2\bar{x}\,n\frac{1}{n}\sum x_i + n\bar{x}^2\right]$$
$$= \frac{1}{n-1}\left[\sum x_i^2 - 2n\bar{x}^2 + n\bar{x}^2\right]$$
$$= \frac{1}{n-1}\left[\sum x_i^2 - n\bar{x}^2\right]$$
$$= \frac{1}{n-1}\left[\sum x_i^2 - n\left(\frac{1}{n}\sum x_i\right)^2\right]$$
$$= \frac{1}{n-1}\left[\sum x_i^2 - \frac{1}{n}\left(\sum x_i\right)^2\right]$$

3.55 a. $s_x^2 = \frac{1}{7}(1210.8055 - \frac{1}{8}(9147.2009)) = 9.6293$

$s_x = \sqrt{9.6293} = 3.1031$

b. $s_y^2 = \frac{1}{7}(59329.4675 - \frac{1}{8}(448212.8432)) = 471.8374$

$s_y = \sqrt{471.8374} = 21.7218$

c. $s_y^2 = (7^2)s_x^2$; $s_y = 7s_x$

d. $s_y^2 = a^2 s_x^2$; $s_y = |a|s_x$

3.57 No. The subset with the smallest 7 numbers has a sample mean $\bar{x} = 5.1429$, which is greater

than 5. Any other subset will have a sample mean greater than 5.1429.

3.59 Answers will vary. Larger values of g_1 indicate a greater lack of symmetry, or more skewness. Larger values (in magnitude) of g_2 suggest a more defined peak in the distribution. Smaller values of g_2 suggest a flatter, more uniform distribution.

Section 3.3: The Empirical Rule and Measures of Relative Standing

3.61 a. $(15, 25)$; $(10, 30)$; $(5, 35)$

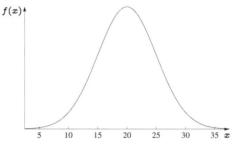

b. $(36.8, 37.2)$; $(36.6, 37.4)$; $(36.4, 37.6)$

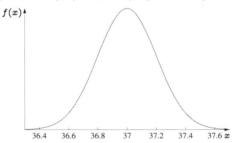

c. $(425, 925)$; $(175, 1175)$; $(-75, 1425)$

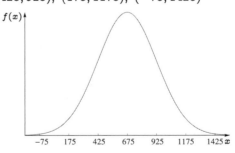

d. $(-17.5, 6.5)$; $(-29.5, 18.5)$; $(-41.5, 30.5)$

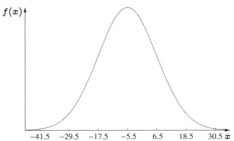

e. $(96.9, 100.3)$; $(95.2, 102.0)$; $(93.5, 103.7)$

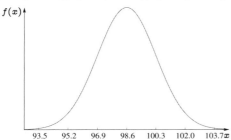

f. $(5130, 5430)$; $(4980, 5580)$; $(4830, 5730)$

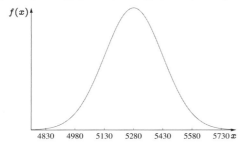

3.63 a. $2.3 = \frac{x-25}{5} \Rightarrow x = 36.5$

b. $-0.7 = \frac{x-9.8}{1.2} \Rightarrow x = 8.96$

c. $1.25 = \frac{x-(-456)}{37} \Rightarrow x = -409.75$

d. $-1.96 = \frac{x-37.6}{5.9} \Rightarrow x = 26.036$

e. $3.5 = \frac{x-55}{0.05} \Rightarrow x = 55.175$

f. $1.28 = \frac{x-3.14}{0.5} \Rightarrow x = 3.78$

g. $-2.5 = \frac{x-2.35}{0.94} \Rightarrow x = 0.0000$

h. $0.55 = \frac{x-0.529}{1.9} \Rightarrow x = 1.574$

3.65 a. $\bar{x} \pm s = 26.5 \pm 4.3 = (22.2, 30.8)$;
$\bar{x} \pm 2s = 26.5 \pm 2(4.3) = (17.9, 35.1)$

b. $(17.9, 35.1)$ is a symmetric interval about the mean, $k = 2$ standard deviations in each direction. Using Chebyshev's Rule, at least $1 - (1/2^2) = 0.75$ of the times lie in this interval.

3.67 85% of all fish caught in the tournament weighed the same or less than the one caught by Ruskey, and 15% weighed more.

3.69 a. $\bar{x} \pm 2s = 1125 \pm 2(250) = (625, 1625)$ is a symmetric interval about the mean, $k = 2$ standard deviations in each direction. Using Chebyshev's Rule, at least $1 - (1/2^2) = 0.75$ of the areas lie in this interval.

b. $\bar{x} \pm 3s = 1125 \pm 3(250) = (375, 1875)$ is a symmetric interval about the mean, $k = 3$ standard deviations in each direction. Using Chebyshev's Rule, at least $1 - (1/3^2) = 0.8889$ of the areas lie in this interval.

c. Form part (c), at least 0.8889 of the areas lie in the interval $(375, 1875)$. Therefore, at most 0.1111 of the areas lie outside this interval. And, at most 0.1111 of the area is less than 375 (cannot assume the distribution is symmetric).

d. $\bar{x} \pm 1.5s = 1125 \pm (1.5)(250) = (750, 1500)$ is a symmetric interval about the mean, $k = 1.5$ standard deviations in each direction. Using Chebyshev's Rule, at least $1 - (1/1.5^2) = 0.5556$ of the areas lie in this interval.

3.71 Consider the following number line.

$$
\begin{array}{ccccccc}
22.9 & 23.8 & 24.7 & 25.6 & 26.5 & 27.4 & 28.3
\end{array}
$$

$$
\bar{x} - 3s \quad \bar{x} - 2s \quad \bar{x} - s \quad \bar{x} \quad \bar{x} + s \quad \bar{x} + 2s \quad \bar{x} + 3s
$$

a. $(23.8, 27.4)$ is a symmetric interval about the mean, 2 standard deviations in each direction. Using the Empirical Rule, approximately 0.95 of the observations lie in this interval.

b. Between 22.9 and 25.6: $0.997/3 = 0.4985$

Between 25.6 and 26.5: $0.68/2 = 0.34$

Between 22.9 and 26.5: $0.4985 + 0.34 = 0.8385$

c. Between 25.6 and 27.4: $0.95/2 = 0.475$

Less than 25.6: 0.5

Less than 27.4: $0.5 + 0.475 = 0.975$

3.73 $z_1 = \frac{33-37}{5} = -0.8$; $z_2 = \frac{35-42}{7} = -1.0$

The second service actually performed *better*. The second service had a time that was farther away from the mean to the left in standard deviations.

3.75 a. Frequency histogram:

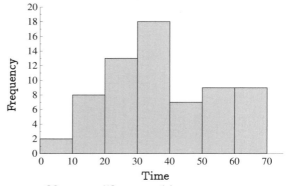

b. $p_{45} \approx 32$, $p_{80} \approx 53$, $p_{10} \approx 14$

c. $d_{45} = (66)(0.45) = 29.7 \Rightarrow$ depth $= 30$; $p_{45} = 32$

$d_{80} = (66)(0.80) = 52.8 \Rightarrow$ depth $= 53$; $p_{80} = 53$

$d_{10} = (66)(0.10) = 6.6 \Rightarrow$ depth $= 7$; $p_{10} = 14$

3.77 a. $\bar{x} = 12.9407$; $s = 1.1382$

Interval	Proportion
Within $1s$: $(11.8025, 14.0789)$	0.6333
Within $2s$: $(10.6644, 15.2171)$	0.9667
Within $3s$: $(\ 9.5262, 16.3553)$	1.0000

b. These proportions are close to the Empirical Rule proportions. There is no evidence to suggest the distribution of times is nonnormal.

c. Frequency histogram:

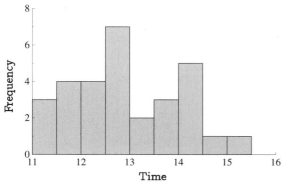

The shape of the distribution appears to be approximately bimodal. It is slightly skewed to the right.

3.79 a. $\bar{x} = \frac{1}{15}(46.5 + \cdots + 42.4) = 40.3867$

$s^2 = \frac{1}{14}(24623.66 - \frac{1}{15}(366993.64)) = 11.2441$

$s = \sqrt{11.2441} = 3.3532$

b. $\sum z_i^2 = [(1.8231)^2 + (-0.6223)^2 + \cdots + (0.6004)^2]$

$\qquad = 3.3238 + 0.3872 + \cdots + 0.3605 = 14$

c. $\sum z_i^2 = \sum \left(\frac{x_i - \bar{x}}{s}\right)^2 = \frac{1}{s^2}\sum(x_i - \bar{x})^2$

$\qquad = \frac{1}{s^2}(n-1)s^2 = n - 1$

Section 3.4: Five-Number Summary and Box Plots

3.81 a. $x_{\min} = 28.0$, $Q_1 = 32.0$, $\tilde{x} = 34.5$, $Q_3 = 35.0$, $x_{\max} = 40.0$

b. $x_{\min} = 52.0$, $Q_1 = 57.0$, $\tilde{x} = 66.5$, $Q_3 = 70.5$, $x_{\max} = 78.0$

c. $x_{\min} = 80.0$, $Q_1 = 83.0$, $\tilde{x} = 91.5$, $Q_3 = 94.0$, $x_{\max} = 98.0$

d. $x_{\min} = 0.4$, $Q_1 = 1.0$, $\tilde{x} = 1.95$, $Q_3 = 2.4$, $x_{\max} = 10.9$

e. $x_{\min} = 103.1$, $Q_1 = 119.9$, $\tilde{x} = 141.9$, $Q_3 = 159.7$, $x_{\max} = 196.9$

f. $x_{\min} = -40.1$, $Q_1 = -33.8$, $\tilde{x} = -28.0$, $Q_3 = -18.5$, $x_{\max} = -9.8$

3.83

	IQR	IF_L	IF_H	OF_L	OF_H
a.	24.0	-14.0	82.0	-50.0	118.0
b.	51.0	1178.5	1382.5	1102.0	1459.0
c.	9.46	51.56	89.4	37.37	103.59
d.	225.6	576.5	1478.9	238.1	1817.3
e.	2.795	-2.9175	8.2625	-7.11	12.455
f.	2.245	-3.1025	5.8775	-6.47	9.245
g.	9.68	-48.19	-9.38	-62.71	5.14
h.	0.38	97.86	99.38	97.29	99.95

3.85

	$x_{\min}$	Q_1	$\tilde{x}$	Q_3	$x_{\max}$
a.	20.0	40.0	55.0	60.0	85.0
b.	-1.5	1.4	1.9	2.8	3.9
c.	4.5	5.5	6.3	7.2	9.5
d.	75.0	93.0	103.0	109.0	119.0
e.	0.0	0.8	1.6	2.9	9.2
f.	0.0	2.5	5.5	9.0	34
g.	-80.0	-58.0	-51.0	-45.0	-22.0

3.87

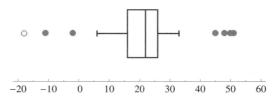

The distribution is centered near 22, slightly negatively skewed, lots of variability, 6 mild outliers and 1 extreme outlier.

3.89 The distribution is negatively skewed, centered near 3.4, with little variability.

3.91

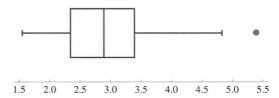

The distribution is slightly positively skewed, centered near 2.8, has little variability, and 1 mild outlier.

3.93

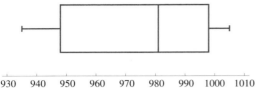

The distribution is slightly negatively skewed, centered near 981, has little variability, and there are no outliers. A standard box plot would look the same since there are no outliers.

3.95 a.

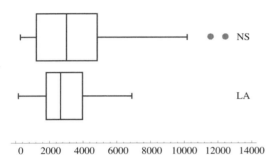

b. The natural science data are centered slightly higher than the liberal arts data, have more variability, are positively skewed, and have 3 mild outliers.

c. The graphs suggest that on average, the natural science faculty use the copier more than the liberal arts faculty.

3.97 a.

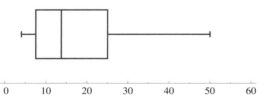

b. The distribution is centered near 14, has lots of variability, is positively skewed, and has no outliers.

c.

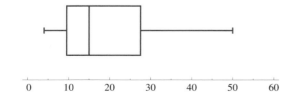

The new box plot looks nearly the same.

Chapter Exercises

3.99 a. $\bar{x} = \frac{1}{20}(156 + \cdots + 173) = 177.35$

$s^2 = \frac{1}{19}(633003 - \frac{1}{20}(12581209)) = 207.5026$

$s = \sqrt{207.5026} = 14.4050$

b.

Interval	Proportion
Within $1s$: $(162.945, 191.755)$	0.70
Within $2s$: $(148.540, 206.160)$	1.00
Within $3s$: $(134.135, 220.565)$	1.00

c. Since these proportions are close to the Empirical Rule proportions, there is no evidence to suggest the distribution is nonnormal.

3.101 a. $\tilde{x} = \frac{1}{2}(236 + 251) = 243.5$

$d_1 = 26/4 = 6.5 \Rightarrow \text{depth} = 7; \ Q_1 = 200$

$d_3 = 3(26)/4 = 19.5 \Rightarrow \text{depth} = 20; \ Q_3 = 361$

$IQR = 361 - 200 = 161$

b. $d_{30} = (26)(0.30) = 7.8 \Rightarrow \text{depth} = 8; \ p_{30} = 201$

$d_{45} = (26)(0.95) = 24.7 \Rightarrow \text{depth} = 25; \ p_{95} = 588$

c. $p_{93} = 438$, therefore, 438 lies in the 93rd percentile.

3.103 a.

	R	s^2	IQR	CV	CQV
Over	5.4000	1.8023	1.7000	2.0707	1.3127
Mid	10.1000	7.9009	4.3000	4.6919	0.0361

b. The summary statistics in part (a) suggest the mid–over racket tensions have more variability.

c.

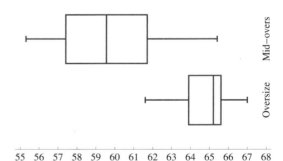

The box plots also suggest the mid–over racket tensions have more variability.

3.105 a. $\bar{x} = \frac{1}{18}(89 + \cdots + 89) = 97.1111$

$\tilde{x} = \frac{1}{2}(96 + 97) = 96.5$

$s^2 = \frac{1}{17}(172412 - \frac{1}{18}(3055504)) = 156.5752$

$s = \sqrt{156.5752} = 12.5130$

b.

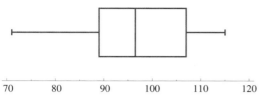

c. The summary statistics and the box plot suggest the distribution is approximately symmetric. The distribution is centered around 97, has lots of variability, and no outliers.

d. A person who drinks three cups of coffee has, on average, around 291 mg ($= 3 \times 97$) of caffeine. This is under the moderate amount of 300 mg.

3.107 a. Standard group:
$$\bar{x} = \tfrac{1}{22}(7.5 + \cdots + 8.6) = 7.6$$
$$s_x^2 = \tfrac{1}{21}(1279.02 - \tfrac{1}{22}(27955.84)) = 0.3952$$
$$s_x = \sqrt{0.3952} = 0.6287$$

Whole group:
$$\bar{y} = \tfrac{1}{25}(6.9 + \cdots + 8.5) = 7.376$$
$$s_y^2 = \tfrac{1}{24}(1451.3 - \tfrac{1}{25}(34003.36)) = 3.7986$$
$$s_y = \sqrt{3.7986} = 1.9490$$

b.

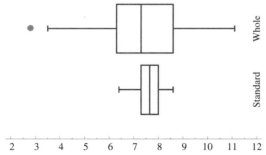

c. The whole language reading speeds have much more variability and the center of the distribution is slightly smaller.

3.109 a.

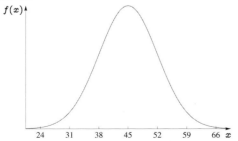

b. A tire harness of 30 is unusually soft since 30 is over 2 standard deviations below the mean.

c. Approximately 0.84 of the tire harnesses are at or below 52.

3.111 Consider the following number line.

1283	2533	3783	5033	6283	7533	8783

$\bar{x} - 3s \quad \bar{x} - 2s \quad \bar{x} - s \quad \bar{x} \quad \bar{x} + s \quad \bar{x} + 2s \quad \bar{x} + 3s$

a. $(3783, 6283)$ is a symmetric interval about the mean, 1 standard deviation in each direction. Using the Empirical Rule, approximately 0.68 of the observations lie in this interval

b. Using the Empirical Rule, there are approximately $1 - 0.997 = 0.003$ of the observations outside the interval $(1283, 8783)$. Therefore, there are approximately $0.003/2 = 0.0015$ of the observations less than 1283.

c. Using the Empirical Rule, there are approximately $0.997/2 = 0.4985$ of the observations in the interval $(5033, 8783)$.

d. This observations, 6515, is within 2 standard deviations of the mean. Therefore it is a reasonable observation subject to normal variation. There is no reason to believe the *average* depth has changed.

3.113 a. $\bar{x} \pm 2s = 3.53 \pm 2(0.1) = (3.33, 3.73)$. Using Chebyshev's Rule, at least $1 - 1/2^2 = 0.75$ of the jelly beans have volume between 3.33 and 3.73.

$\bar{x} \pm 3s = 3.53 \pm 3(0.1) = (3.23, 3.83)$. Using Chebyshev's Rule, at least $1 - 1/3^2 = 0.8889$ of the jelly bean volumes are between 3.23 and 3.83. Therefore, at most $1 - 0.8889 = 0.1111$ have volume less than 3.23.

b. Claim: $\mu = 3.53$
Experiment: $x = 3.1$
Likelihood: 3.1 is over 4 standard deviations from the mean. This is a very unlikely observation.
Conclusion: There is evidence to suggest the manufacturer's claim is false.

3.115 a. $z = \frac{1-0.7}{0.1} = 3.0$. It is unlikely a fisherman will catch a small mouth bass with mercury level greater than 1 because this is 3 standard deviations from the mean.

b. $z = \frac{1-0.7}{0.05} = 6$. It is even more unlikely a fisherman will catch a small mouth bass with mercury level greater than 1 because this is 6 standard deviations from the mean.

c.

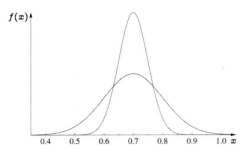

Chapter 4: Probability

Section 4.1: Experiments, Sample Spaces, and Events

4.1

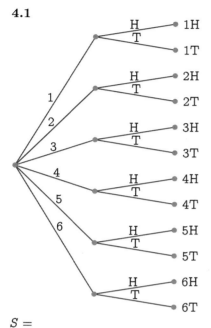

$S =$
$\{1H, 2H, 3H, 4H, 5H, 6H, 1T, 2T, 3T, 4T, 5T, 6T\}$

4.3 There are $5 \times 5 = 25$ outcomes.

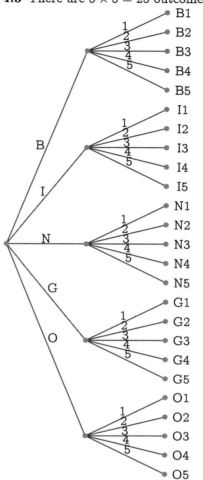

4.5 a. $A' = \{1, 3, 5, 7, 9\}$
b. $C' = \{5, 6, 7, 8, 9\}$
c. $D' = \{0, 1, 2, 3, 4\}$
d. $A \cup B = \{0, 1, 2, 3, 4, 5, 6, 7, 8, 9\} = S$
e. $A \cup C = \{0, 1, 2, 3, 4, 6, 8\}$
f. $A \cup D = \{0, 2, 4, 5, 6, 7, 8, 9\}$

4.7 a. $A' = \{b, d, f, h, i, j, k\}$
b. $C' = \{a, b, d, e, j, k\}$
c. $D' = \{c, f, i\}$
d. $A \cap B = \{c\}$
e. $A \cap C = \{c, g\}$
f. $C \cap D = \{g, h\}$

4.9 a. $(A \cap B \cap C)' = \{a, b, d, e, f, g, h, i, j, k\}$
b. $A \cup B \cup C \cup D = \{a, b, c, d, e, f, g, h, i, j, k\}$
c. $(B \cup C \cup D)' = \{\ \}$
d. $B' \cap C' \cap D' = \{\ \}$

4.11

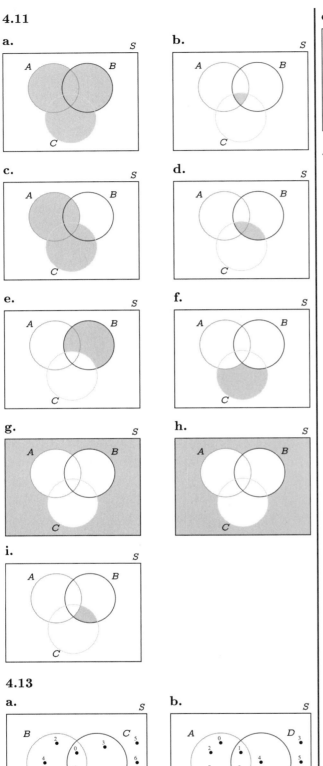

4.13

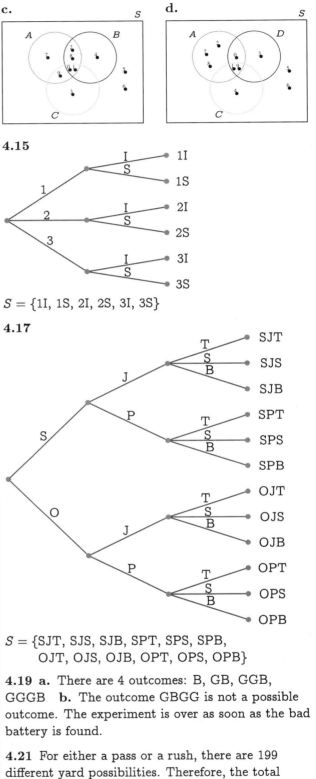

4.15

$S = \{1I, 1S, 2I, 2S, 3I, 3S\}$

4.17

$S = \{$SJT, SJS, SJB, SPT, SPS, SPB,
OJT, OJS, OJB, OPT, OPS, OPB$\}$

4.19 a. There are 4 outcomes: B, GB, GGB,
GGGB   **b.** The outcome GBGG is not a possible
outcome. The experiment is over as soon as the bad
battery is found.

4.21 For either a pass or a rush, there are 199
different yard possibilities. Therefore, the total
number of outcomes is $n = 2 \times 199 = 398$.

4.23 a. $S = \{$LS, LU, LV, LP, RS, RU, RV, RP, SS,
SU, SV, SP$\}$

b. $A = \{LV, RV, SV\}$,
$B = \{LS, LP, RS, RP, SS, SP\}$,
$C = \{LS, LU, LV, LP\}$,
$D = \{RS, RU, RV, RP, SS, SU, SV, SP\}$
c. $C \cup D = S$, $C \cap D = \{\ \}$

4.25 a. $S =$
$\{A0, A1, A2, A3, A4, A5, F0, F1, F2, F3, F4, F5\}$
b. A = The passenger has 0 bags.
B = The passenger is foreign.
C = The passenger has 1 or 2 bags.
D = The passenger is foreign and has 0 or 5 bags.
E = The passenger has an odd number of bags.

4.27 a.

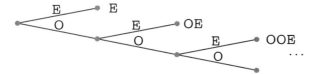

b. $S = \{E, OE, OOE, OOOE, \ldots\}$

4.29 a. $S = \{R1, R2, R3, R4, R5, J1, J2, J3, J4,$
$J5, N1, N2, N3, N4, N5, C1, C2, C3, C4, C5\}$
b. $A' = \{C1, C2, C3, C4, C5, J1, J2, J3, J4, J5, N1,$
$N2, N3, N4, N5\}$
$A \cup C = \{C1, C2, J1, J2, N1, N2, R1, R2, R3, R4,$
$R5\}$
$A \cap D = \{\ \}$
$C \cap D = \{C1\}$
$A \cap C \cap D = \{\ \}$
$A \cap B = \{\ \}$

4.31
a. $S = \{B0, B1, B2, B3, B4, P0, P1, P2, P3, P4\}$
b. $A \cup B = \{B0, B1, B2, B3, B4, P0\}$
$A \cap B = \{B0\}$
$B \cup C = \{B0, B1, P0, P1\}$
$B \cap C = \{B0, P0\}$
$A \cap D = \{B3\}$
$A \cap B \cap C \cap D = \{\ \}$

Section 4.2: An Introduction to Probability

4.33 a. $P(A) = P(e_1) + P(e_2) + P(e_3)$
$= 0.07 + 0.09 + 0.13 = 0.29$
b. $P(C) = P(e_1) + P(e_5) + P(e_7)$
$= 0.07 + 0.22 + 0.16 = 0.45$
c. $P(d) = P(e_3) + P(e_4) + P(e_5) + P(e_6) + P(e_7)$
$= 0.13 + 0.18 + 0.22 + 0.15 + 0.16 = 0.84$
d. $P(A \cup B) = P(\{e_1, e_2, e_3, e_4, e_6, e_7\})$
$= 1 - P(e_5) = 1 - 0.22 = 0.78$

e. $P(A \cap C) = P(e_1) = 0.07$
f. $P(B \cap D) = P(\{e_4, e_6, e_7\})$
$= P(e_4) + P(e_6) + P(e_7)$
$= 0.18 + 0.15 + 0.16 = 0.49$
g. $P(A') = 1 - P(A) = 1 - 0.29 = 0.71$
h. $P(A \cap C') = P(\{e_2, e_3\})$
$= 0.09 + 0.13 = 0.22$
i. $P(A' \cap D) = P(\{e_4, e_5, e_6, e_7\})$
$= P(e_4) + P(e_5) + P(e_6) + P(e_7)$
$= 0.18 + 0.22 + 0.15 + 0.16 = 0.71$
j. $P(C') = 1 - P(C) = 1 - 0.45 = 0.55$
k. $P(B \cap C \cap D) = P(e_7) = 0.16$
l. $P[(B \cup C)'] = P(e_3) = 0.13$

There are no other possible simple events in this experiment since the probabilities sum to 1.

4.35 a. $P(A) = \frac{6}{22} = 0.2727$
b. $P(B) = \frac{11}{22} = 0.5$
c. $P(C) = \frac{11}{22} = 0.5$
d. $P(D) = \frac{3}{22} = 0.1364$

4.37 a. $P(A \cap B) = P(A) + P(B) - P(A \cup B)$
$= 0.26 + 0.68 - 0.80 = 0.14$
b. $P(A') = 1 - P(A) = 1 - 0.26 = 0.74$
c. $P[(A \cap B)'] = 1 - P(A \cap B)$
$= 1 - 0.14 = 0.86$
d. $P[(A \cup B)'] = 1 - P(A \cup B)$
$= 1 - 0.80 = 0.20$

4.39

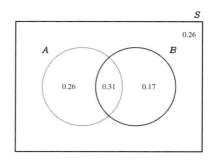

4.41 a. $n = \frac{5 \cdot 4 \cdot 3}{\text{Number of repeats}} = \frac{60}{6} = 10$
$\{HLP, HLD, HLV, HPD, HPV, HDV, LPD, LPV,$
$LDV, PDV\}$
b. P(includes health insurance) $= \frac{6}{10} = 0.60$
c. P(life insurance and prescription plan) $= \frac{3}{10} = 0.30$

4.43 a. $P(1\text{–}10) = 0.141 + 0.161 + 0.289 = 0.591$

b. P(life or death) $= 0.032 + 0.001 = 0.033$

c. P(not death) $= 1 - $ P(death) $= 1 - 0.001 = 0.999$

4.45 a. P(state or local law prohibition)
$= 0.089 + 0.001 = 0.090$

b. P(not domestic violence)
$= 1 - (0.067 + 0.036) = 1 - 0.103 = 0.897$

c. P(neither felony nor other)
$= 1 - (0.386 + 0.317) = 1 - 0.703 = 0.297$

4.47 a. $P(A) = 0.08 + 0.15 = 0.23$
$P(B) = 0.30 + 0.10 + 0.12 = 0.52$
$P(C) = 0.08 + 0.15 + 0.12 + 0.05 = 0.40$

b. $P(A \cup B)$
$= $ P({magazine, newspaper, TV, radio, Internet})
$= $ P(A) + P(B) $= 0.23 + 0.52 = 0.75$
$P(A \cap B) = $ P({ }) $= 0$
$P(B \cap C) = $ P(Internet) $= 0.12$

c. $P(A') = 1 - $ P(A) $= 1 - 0.23 = 0.77$
$P(A' \cap C) = $ P({Internet, billboard})
$= 0.12 + 0.05 = 0.17$
$P(A \cap B \cap C) = $ P({ }) $= 0$

d. $P(B' \cap C') = $ P({not seen}) $= 0.20$
$P[(B \cup C)'] = $ P({not seen}) $= 0.20$

4.49 a. $P(A) = 0.479 + 0.030 = 0.509$

$P(B) = 1 - 0.345 = 0.655$

$P(C) = 0.133$

b. $P(A \cap B)$
$= $ P({utility gas, bottled, tank, or LP gas})
$= 0.479 + 0.030 = 0.509$

$P(A \cup C) = $
P({utility gas, bottled, tank, or LP gas, electricity})
$= 0.479 + 0.030 + 0.133 = 0.642$

$P(A \cap C) = $ P({ }) $= 0$

c. $P(A' \cup C)$
$= $ P({electricity, fuel oil, kerosene, etc.,
        coal or coke, wood,
        other fuel, no fuel used})
$= 0.133 + 0.345 + 0.001 + 0.006 + 0.004 + 0.002 = 0.491$

$P(A \cup B \cup C') = $ P(S) $= 1$

4.51 a.

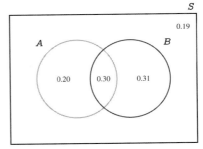

b. $P(A \cup B) = $ P(A) + P(B) $-$ P(A $\cap$ B)
            $= 0.50 + 0.61 - 0.30 = 0.81$

c. $P[(A \cup B)'] = 1 - $ P(A $\cup$ B) $= 1 - 0.81 = 0.19$

d. P(only helmet) $= $ P(A) $-$ P(A $\cap$ B) $=$
$0.50 - 0.30 = 0.20$

4.53 a. P(positive)
$= 0.374 + 0.357 + 0.085 + 0.034 = 0.850$

b. P(type B) $= 0.085 + 0.015 = 0.10$

c. P(not type O)
$= 1 - $ P(type O)
$= 1 - (0.374 + 0.066) = 1 - 0.44 = 0.56$

d. P(type AB or negative)
$= 0.034 + 0.066 + 0.063 + 0.015 + 0.006 = 0.184$
$= $ P(type AB) $+$ P(negative) $-$ P(type AB$\cap$negative)

4.55 a.

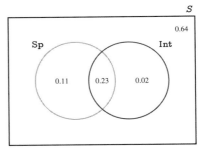

b. P(Sp $\cup$ Int) $= $ P(Sp) + P(Int) $-$ P(Sp $\cap$ Int)
$= 0.34 + 0.25 - 0.23 = 0.36$

c. $P[(Sp \cup Int)']$
$= 1 - $ P(Sp $\cup$ Int) $= 1 - 0.36 = 0.64$

d. P(Sp $\cup$ Int) $-$ P(Sp $\cap$ Int)
$= 0.36 - 0.23 = 0.13$

Section 4.3: Counting Techniques

4.57 a. $_8P_4 = \frac{8!}{4!} = 1680$

b. $_{11}P_7 = \frac{11!}{4!} = 1,663,200$ **c.** $_{12}P_4 = \frac{12!}{8!} = 11,880$

d. $_{10}P_{10} = \frac{10!}{0!} = 3,628,800$ **e.** $_{10}P_1 = \frac{10!}{9!} = 10$

f. $_{10}P_0 = \frac{10!}{10!} = 1$ **g.** $_9P_2 = \frac{9!}{7!} = 72$

h. $_{20}P_2 = \frac{20!}{18!} = 380$ **i.** $_{100}P_2 = \frac{100!}{98!} = 9900$

4.59 $9! = 362,880$

4.61 $_{20}P_7 = \frac{20!}{13!} = 390{,}700{,}800$

4.63 $n = {}_{20}P_3 = \frac{20!}{17!} = 6840$

4.65 a. $n = 40 \times 40 \times 40 = 64{,}000$

b. P(single-digit numbers) $= \frac{1000}{64000} = 0.0156$

c. $n = {}_{40}P_3 = \frac{40!}{3!\,37!} = 40 \times 39 \times 38 = 59{,}280$

P(single-digit numbers) $= \frac{720}{59280} = 0.0121$

4.67 a. Number of policies $= 4 \cdot 2 \cdot 3 \cdot 3 \cdot 3 = 216$

b. Number of polices with comprehensive coverage of at least \$500,000 $= 4 \cdot 2 \cdot 3 \cdot 3 \cdot 2 = 144$

c. Number of policies with bodily injury liability and property damage liability of \$100,000 $= 4 \cdot 1 \cdot 1 \cdot 3 \cdot 3 = 36$

4.69 a. Number of outcomes in the sample space $= \binom{15}{4} = 1365$

P(4 defect-free) $= \binom{12}{4}/1365 = 495/1365 = 0.3626$

b. P(3 defective) $= \frac{\binom{12}{1}\binom{3}{3}}{1365} = \frac{12 \cdot 1}{1365} = 0.0088$

c. P(at least 1 defective)
$= 1 - \text{P(0 defective)}$
$= 1 - \text{P(4 defect-free)}$
$= 1 - 0.3626 = 0.6374$

4.71 a. Number of calling schedules
$= {}_{12}P_8 = 19{,}958{,}400$

b. P(2 who will purchase are first 2 called)
$= \frac{({}_2P_2)({}_{10}P_6)}{19{,}958{,}400} = 0.0152$

c. P(2 on do-not-call list not called)
$= {}_{10}P_8/19{,}958{,}400 = 0.0909$

4.73 a. Number of ways to select 3 songs
$= \binom{15}{3} = 455$

b. P(all 3 selected are female vocalists)
$= \binom{5}{3}/455 = 10/455 = 0.022$

c. P(0 female vocalists)
$= \binom{10}{3}/455 = 120/455 = 0.2637$

d. P(explicit lyrics song not selected)
$= \binom{14}{3}/455 = 364/455 = 0.80$

4.75 a. Number of ways the bids can be opened
$= {}_8P_8 = 40320$

b. P(lowest bid opened first)
$= 1 \cdot {}_7P_7/40320 = 5040/40320 = 0.125$

4.77 a. Number of different flight crews
$= (10)(15)(17) = 2550$

b. P(2 individuals not on flight crew)
$= (9)(14)(17)/2550 = 2142/2550 = 0.84$

c. P(flight crew includes qualified navigator)
$= (10)(15)(8)/2550 = 1200/2550 = 0.4706$

4.79 a. Number of wash-dry crews $= {}_5P_2 = 20$

b. P(two girls selected)
$= {}_2P_2/20 = 2/20 = 0.10.$

Since this probability is small, there is evidence to suggest the selection process was not random.

4.81 a. $n = {}_{10}P_{10} = 10! = 3{,}628{,}800$

b. P(twins in the middle)
$\frac{8 \cdot 7 \cdot 6 \cdot 5 \cdot 2 \cdot 1 \cdot 4 \cdot 3 \cdot 2 \cdot 1}{3{,}628{,}800} = \frac{80{,}640}{3{,}628{,}800} = 0.0222$

c. P(twins side-by-side)
$= \frac{9 \cdot 80640}{3{,}628{,}800} = \frac{725{,}760}{3{,}628{,}800} = 0.20$

d. P(picture arrangement alternates)
$= (2)\frac{5 \cdot 5 \cdot 4 \cdot 4 \cdot 3 \cdot 3 \cdot 2 \cdot 2 \cdot 1 \cdot 1}{3{,}628{,}800} = \frac{28{,}800}{3{,}628{,}800} = 0.0079$

4.83 a. Number of pre-flop hands $= \binom{52}{2} = 1326$

b. P(pre-flop hand consists of two aces)
$= \binom{4}{2}/1326 = 6/1326 = 0.0045$

c. P(pre-flop hand consists of a pair)
$= \frac{13 \cdot \binom{4}{2}}{1326} = \frac{78}{1326} = 0.0588$

d. P(pre-flop hand same suit)
$= \frac{4 \cdot \binom{13}{2}}{1326} = \frac{4 \cdot 78}{1326} = \frac{312}{1326} = 0.2353$

4.85 a. sum $= 1 + 2 + 1 = 4$

b. sum $= 1 + 3 + 3 + 1 = 8$

c. sum $= 1 + 4 + 6 + 4 + 1 = 16$

d. sum $= 1 + n + \cdots + n + 1 = 2^n$

4.87 Number of ways to arrange n people at a round table $= n!/n = (n-1)!$

4.89 a. $\binom{100}{22} = 7.3321 \times 10^{21}$
$(= 7{,}332{,}066{,}885{,}177{,}656{,}269{,}200)$

b. P(all Democrats)
$= \frac{\binom{45}{22}}{\binom{100}{22}} = 5.61467 \times 10^{-10} = 0.000000000561467$

c. P(more Democrats than Republicans)
$= \text{P(12 Dems)} + \text{P(13 Dems)} + \cdots + \text{P(22 Dems)}$
$= \frac{\binom{45}{12}\binom{55}{10}}{\binom{100}{22}} + \cdots + \frac{\binom{45}{22}\binom{55}{0}}{\binom{100}{22}}$
$= 0.2185$

Section 4.4: Conditional Probability

4.91 a. Conditional. **b.** Unconditional.
c. Unconditional. **d.** Conditional.
e. Unconditional.

4.93

	B_1	B_2	B_3	
A_1	0.095	0.016	0.007	0.118
A_2	0.205	0.188	0.003	0.396
A_3	0.155	0.238	0.093	0.486
	0.455	0.442	0.103	1.000

a. $P(A_1) = 0.095 + 0.016 + 0.007 = 0.118$

$P(A_2) = 0.205 + 0.188 + 0.003 = 0.396$

$P(A_3) = 0.155 + 0.238 + 0.093 = 0.486$

b. $P(B_1) = 0.095 + 0.205 + 0.155 = 0.455$

$P(B_2) = 0.016 + 0.188 + 0.238 = 0.442$

$P(B_3) = 0.007 + 0.003 + 0.093 = 0.103$

c. $P(A_1 \cap B_1) = 0.095$

$P(A_2 \cap B_2) = 0.188$

$P(A_3 \cap B_3) = 0.093$

d. $P(A_1 \mid B_1) = \frac{0.095}{0.455} = 0.2088$

e. $P(B_1 \mid A_1) = \frac{0.095}{0.118} = 0.8051$

$P(A_1' \cap B_1') = 0.188 + 0.003 + 0.238 + 0.093 = 0.5220$

f. $P(B_2 \mid A_2) = \frac{0.188}{0.396} = 0.4747$

$P(B_3 \mid A_3) = \frac{0.093}{0.486} = 0.1914$

4.95 a.

	B_1	B_2	B_3	
A_1	178	231	406	815
A_2	123	150	244	517
A_3	165	202	335	702
	466	583	985	2034

b. Number of people who participated in the survey is the total of the last column and also the total of the last row, 2034.

c. $P(A_1) = \frac{815}{2034} = 0.4007$

$P(A_2) = \frac{517}{2034} = 0.2542$

$P(A_3) = \frac{702}{2034} = 0.3451$

d. $P(B_1 \cap A_1) = \frac{178}{2034} = 0.0875$

$P(B_2 \cap A_2) = \frac{150}{2034} = 0.0737$

$P(B_3 \cap A_3) = \frac{335}{2034} = 0.1647$

e. $P(A_3 \mid B_1) = \frac{165}{466} = 0.3541$

$P(B_2 \mid A_2) = \frac{150}{517} = 0.2901$

$P(A_3 \cap B_3') = \frac{165+202}{2034} = 0.1804$

4.97 a. $P(A) = 0.574$; $P(B) = 0.488$; $P(C) = 0.465$

b. $P(A \cap B) = P(2) + P(5) + P(6)$
$$= 0.142 + 0.083 + 0.072 = 0.297$$

$P(B \cap C) = P(5) + P(6) + P(7)$
$$= 0.083 + 0.072 + 0.063 = 0.218$$

c. $P(A \mid B) = \frac{0.297}{0.488} = 0.6086$

$P(B \mid C) = \frac{0.218}{0.465} = 0.4688$

$P[(A \cap B) \mid C] = \frac{0.155}{0.465} = 0.3333$

d. $P(0 \mid C') = \frac{0.135}{0.535} = 0.2523$

$P(7 \mid C) = \frac{0.063}{0.465} = 0.1355$

$P[(A \cup B) \mid C'] = \frac{0.400}{0.535} = 0.7477$

e. $P(2 \mid B) = \frac{0.142}{0.488} = 0.2910$

$P(3 \mid B) = \frac{0.128}{0.488} = 0.2623$

$P(7 \mid B) = \frac{0.063}{0.488} = 0.1291$

4.99 $P(C \mid O) = \frac{P(C \cap O)}{P(O)} = \frac{0.15}{0.25} = 0.60$

4.101 a. $P(S \mid C) = \frac{P(S \cap C)}{P(C)} = \frac{0.60}{0.80} = 0.75$

b. $P(S \mid C') = \frac{P(S \cap C')}{P(C')} = \frac{0.15}{0.20} = 0.75$

4.103 a. $P(\text{rush}) = \frac{284}{536} = 0.5299$

b. $P(\text{pass} \mid \text{1st}) = \frac{87}{213} = 0.4085$

c. $P(\text{rush} \mid \text{1st or 2nd}) = \frac{207}{377} = 0.5491$

d. $P(\text{3rd} \mid \text{pass}) = \frac{82}{252} = 0.3254$

4.105

a. $P(CM) = 1 - (P(\text{Dial}) + P(\text{ADSL}) + P(DA))$
$$= 1 - (0.813 + 0.096 + 0.001)$$
$$= 1 - 0.91 = 0.09$$

b. $P(\text{Prem} \mid CM) = \frac{P(\text{Prem} \cap CM)}{P(CM)} = \frac{0.009}{0.09} = 0.10$

c. $P(DA \mid CM') = \frac{0.813}{0.91} = 0.8934$

4.107

	Number of drugs		
	≥ 5	< 5	
Men	220	780	1000
Women	252	648	900
	472	1428	1900

a. $P(W \cap \geq 5) = \frac{252}{1900} = 0.1326$

b. $P(\geq 5 \mid M) = \frac{220}{1000} = 0.22$

c. $P(W \mid \geq 5) = \frac{252}{472} = 0.5339$

4.109

		Garden defense			
		Fence	Chemicals	Nothing	
Result	Pests	0.05	0.08	0.34	0.47
Result	No pests	0.30	0.20	0.03	0.53
		0.35	0.28	0.37	1.00

a. $P(\text{nothing} \mid \text{pests}) = \frac{0.34}{0.47} = 0.7234$

b. $P(\text{pests} \mid \text{chemicals}) = \frac{0.08}{0.28} = 0.2857$

c. The gardener most likely used a fence.

P(fence | no pests) is the largest of the three conditional probabilities.

4.111 a.

	Prison	Type of sentence Conditional sentence	Probation	
Crimes of violence	15302	2791	36466	54559
Property crimes	24443	3619	33193	61255
Administration of justice	21412	1026	13635	36073
Other criminal code offenses	6871	671	9940	17482
Criminal code offenses (traffic)	7327	833	6659	14819
Other federal statute	7292	2214	8584	18090
	82647	11154	108477	202278

(left column label: Criminal code)

b. $P(\text{prop crime} \cap \text{cond sent}) = \frac{3619}{202278} = 0.0179$

c. $P(\text{traffic} \mid \text{probation}) = \frac{6659}{108477} = 0.0614$

d. $P(\text{prison} \mid \text{violence}) = \frac{15302}{54559} = 0.2805$

4.113

	Se	Fr	Re	St	Os	Ns	
Ex	200	128	89	17	25	189	648
Tr	57	40	21	5	10	58	191
Fa	51	35	20	4	14	49	173
Hn	40	34	19	6	12	44	155
Ey	23	20	9	3	7	26	88
Ot	2	3	3	2	1	5	16
	373	260	161	37	69	371	1271

(column group header: Victim-shooter relationship; row label: Body part)

a. $P(\text{Ey} \cap \text{Re}) = \frac{9}{1271} = 0.0071$

b. $P(\text{Ex} \mid \text{St}) = \frac{17}{37} = 0.4595$

c. $P(\text{Fr} \mid \text{Hn}) = \frac{34}{155} = 0.2194$

d. $P(\text{Re} \mid \text{Ey}') = \frac{152}{1183} = 0.1285$

Section 4.5: Independence

4.115 a. Dependent. **b.** Dependent.
c. Independent. **d.** Dependent.

4.117
a. $P(A \cap B) = P(A)P(B \mid A) = (0.25)(0.34) = 0.085$;
$P(B' \mid A) = 1 - P(B \mid A) = 1 - 0.34 = 0.66$;
$P(A \cap B') = P(A)P(B' \mid A) = (0.25)(0.66) = 0.165$
b. $P(A \cap B \cap C) = P(A)P(B \mid A)P[C \mid (A \cap B)]$
$= (0.25)(0.34)(0.62) = 0.0527$
$P[C' \mid (A \cap B)] = 1 - P[C \mid (A \cap B)] = 1 - 0.62 = 0.38$

$P(A \cap B \cap C') = P(A)P(B \mid A)P[C' \mid (A \cap B)]$
$= (0.25)(0.34)(0.38) = 0.0323$
c. There is not enough information to determine independence or dependence.

4.119
a. $P(A \cap B) = P(A)P(B \mid A) = (0.40)(0.25) = 0.10$;
$P(A \cap C) = P(A)P(C \mid A) = (0.40)(0.45) = 0.18$;
$P(A \cap D) = P(A)P(D \mid A) = (0.40)(0.30) = 0.12$
b. There is not enough information to determine independence or dependence.
c. $P(B' \mid A) = 1 - P(B \mid A) = 1 - 0.25 = 0.75$
If the event A occurs, only B, C, or D can occur because $P(B \mid A) + P(C \mid A) + P(D \mid A) = 1$

4.121 a. $P(A') = 1 - P(A) = 1 - 0.35 = 0.65$;
$P(C \mid A) = 1 - (P(B \mid A) + P(D \mid A))$
$= 1 - (0.20 + 0.62) = 1 - 0.82 = 0.18$
$P(B \mid A') = 1 - (P(C \mid A') + P(D \mid A'))$
$= 1 - (0.28 + 0.36) = 1 - 0.64 = 0.36$
b. $P(A \cap C) = P(A)P(C \mid A)$
$= (0.35)(0.18) = 0.063$
$P(A' \cap B) = P(A')P(B \mid A')$
$= (0.65)(0.36) = 0.234$
c. $P(D) = P(A \cap D) + P(A' \cap D)$
$= P(A)P(D \mid A) + P(A')P(D \mid A')$
$= (0.35)(0.62) + (0.65)(0.36) = 0.451$

4.123 Let I_i = coal miner i is injured.
a. $P(I_1 \cap I_1) = P(I_1)P(I_2)$
$= (0.000248)(0.000248) = 0.0000000615$
b. $P(I_1' \cap I_2') = P(I_1')P(I_2')$
$= (0.999752)(0.999752) = 0.999504$
c. $P(1 \text{ injury})$
$= P(I_1 \cap I_2') + P(I_1' \cap I_2)$
$= (0.000248)(0.999752) + (0.999752)(0.000248)$
$= 0.00024794 + 0.00024794 = 0.00049588$

4.125 Let K_i = person i dies from a krait bite.
a. $P(K_1 \cap K_2) = P(K_1)pr(K_2) = (0.50)(0.50) = 0.25$
b. $P(1 \text{ person will die from a krait bite})$
$= P(K_1 \cap K_2') + P(K_1' \cap K_2)$
$= P(K_1)P(K_2') + P(K_1')P(K_2)$
$= (0.5)(0.5) + (0.5)(0.5) = 0.5$
c. $P(\text{at most one person dies})$
$= P(0 \text{ people die}) + P(1 \text{ person dies})$
$= P(K_1' \cap K_2') + 0.50$
$= P(K_1')P(K_2') + 0.50$
$= (0.50)(0.50) + 0.50 = 0.25 + 0.50 = 0.75$

4.127 a. P(all four erupt)

$= P(S \cap D \cap F \cap K)$

$= P(S) P(D) P(F) P(K)$

$= (0.9063)(0.7216)(0.7418)(0.5254) = 0.2549$

b. P(none of the four erupt)

$= P(S' \cap D' \cap F' \cap K')$

$= P(S') P(D') P(F') P(K')$

$= (0.0937)(0.2784)(0.2582)(0.4746) = 0.0032$

c. P(at least 1 erupts)

$= 1 - $ P(none of the four erupt)

$= 1 - 0.0032 = 0.9968$

d. All four volcanoes erupted in 2008. The probability of this occurring is 0.2549.

4.129 Let $Y_i = $ well i yields oil.

a. P(all four wells yield oil)

$= P(Y_1 \cap Y_2 \cap Y_3 \cap Y_4)$

$= P(Y_1) P(Y_2) P(Y_3) P(Y_4)$

$= (0.20)(0.20)(0.20)(0.20) = 0.0016$

b. P(1 of the four wells yields oil)

$= P(Y_1 \cap Y_2' \cap Y_3' \cap Y_4') + P(Y_1' \cap Y_2 \cap Y_3' \cap Y_4')$

$\quad + P(Y_1' \cap Y_2' \cap Y_3 \cap Y_4') + P(Y_1' \cap Y_2' \cap Y_3' \cap Y_4)$

$= 4(0.20)(0.80)^3 = 0.4096$

c. P(at least 2 wells yield oil)

$= 1 - (\text{P}[0 \text{ or } 1 \text{ yield oil}])$

$= 1 - (\text{P}[Y_1' \cap Y_2' \cap Y_3' \cap Y_4'] + 0.4096)$

$= 1 - [(0.80)^4 + 0.4096]$

$= 1 - 0.8192 = 0.1808$

4.131 Let $R_i = $ stock i will rise.

a. P(both stocks will rise)

$= P(R_1 \cap R_2) = P(R_1) P(R_2)$

$= (0.42)(0.63) = 0.2646$

b. $P(R_1 \cap R_2') = P(R_1) P(R_2')$

$= (0.42)(0.37) = 0.1554$

c. $P(R_1 \cap R_2) = P(R_1) P(R_2 \mid R_1)$

$= (0.42)(0.81) = 0.3402$

4.133 Let $S_i = $ makes shot i.

a. P(makes both shots)

$= P(S_1 \cap S_2) = P(S_1) P(S_2 \mid S_1)$

$= (0.886)(0.85) = 0.7531$

b. P(misses both shots)

$= P(S_1' \cap S_2') = P(S_1') P(S_2' \mid S_1')$

$= (0.114)(0.05) = 0.0057$

c. P(makes only 1 shot)

$= P(S_1 \cap S_2') + P(S_1' \cap S_2)$

$= P(S_1) P(S_2' \mid S_1) + P(S_1') P(S_2 \mid S_1')$

$= (0.886)(0.15) + (0.114)(0.95)$

$= 0.1329 + 0.1083 = 0.2412$

4.135 a. P(all three angry)

$= P(R \cap D \cap I) = P(R) P(D) P(I)$

$= (0.56)(0.61)(0.66) = 0.2255$

b. P(all three not angry)

$= P(R' \cap D' \cap I') = P(R') P(D') P(I')$

$= (0.44)(0.39)(0.34) = 0.0583$

c. P(1 of the three is angry)

$= P(R \cap D' \cap I') + P(R' \cap D \cap I') + P(R' \cap D' \cap I)$

$= P(R) P(D') P(I') + P(R') P(D) P(I') +$

$P(R') P(D') P(I)$

$= (0.56)(0.39)(0.34) + (0.44)(0.61)(0.34) +$

$(0.44)(0.39)(0.66)$

$= 0.0743 + 0.0913 + 0.1133 = 0.2788$

4.137 a. $P(\$25 \cap A) = P(\$25) P(A \mid \$25)$

$= (0.40)(0.60) = 0.24$

b. $P(A) = P(\$30 \cap A) + P(\$25 \cap A) + P(\$20 \cap A)$

$= P(\$30) P(A \mid \$30) + P(\$25) P(A \mid \$25)$

$\quad + P(\$20) P(A \mid \$20)$

$= (0.5)(0.9) + (0.4)(0.6) + (0.1)(0.05)$

$= 0.45 + 0.24 + 0.005 = 0.695$

c. $P(\$20 \mid A) = P(\$20 \cap A)/P(A)$

$= (0.005)/(0.695) = 0.0072$

4.139 a. $P(\text{Pos} \cap E) = P(E) P(\text{Pos} \mid E)$

$= (0.05)(0.95) = 0.0475$

b. $P(\text{Pos}) = P(\text{Pos} \cap E) + P(\text{Pos} \cap E')$

$= P(E) P(\text{Pos} \mid E) + P(E') P(\text{Pos} \mid E')$

$= (0.05)(0.95) + (0.95)(0.02) = 0.0665$

c. $P(E \mid \text{Pos}) = P(E \cap \text{Pos})/P(\text{Pos})$

$= (0.0475)/(0.0665) = 0.7143$

4.141 Let $H_i = $ missile i hits the plane.

a. P(at least one missile hits)

$= 1 - $ P(0 missiles hit the plane)

$= 1 - P(H_1' \cap H_2' \cap H_3')$

$= 1 - P(H_1') P(H_2') P(H_3')$

$= 1 - (0.20)^3 = 0.9920$

b. P(0 missiles hit the plane)

$= P(H_1' \cap H_2' \cap H_3')$

$= P(H_1') P(H_2') P(H_3')$

$= (0.20)^3 = 0.008$

c. P(at least one missile hits)

$= 1 - $ P(0 missiles hit the plane)

$= 1 - (0.20)^n \geq 0.9999$

$\Rightarrow (0.20)^n \leq 0.0001 \Rightarrow n \ln(0.20) \leq \ln(0.0001)$

$\Rightarrow n \geq 5.7 \Rightarrow n = 6$

4.143 a. $P(L) = 0.366$, $P(T' \mid D) = 0.774$,
$P(T \mid L) = 0.156$, $P[B' \mid (D \cap T)] = 0.545$,
$P[B' \mid (D \cap T')] = 0.622$, $P[B' \mid (L \cap T)] = 0.105$,
$P[B' \mid (L \cap T')] = 0.005$

b. $P(B \cap T \cap D)$
$= P(D)\, P(T \mid D)\, P[B \mid (D \cap T)]$
$= (0.634)(0.226)(0.455) = 0.0652$

c. $P(B')$
$= P(D \cap T \cap B') + P(D \cap T' \cap B')$
$\quad + P(L \cap T \cap B') + (L \cap T' \cap B')$
$= 0.0781 + 0.3052 + 0.0060 + 0.0015 = 0.3909$

d. $P(B \mid T) = P(B \cap T)/P(T)$
$= 0.1163/0.2004 = 0.5803$

4.145 $P(A \mid R) = 0.2830$
$P(B \mid R) = \frac{P(B \cap R)}{P(R)} = \frac{0.0180}{0.0530} = 0.3396$
$P(C \mid R) = \frac{P(C \cap R)}{P(R)} = \frac{0.0200}{0.0530} = 0.3774$
The salesperson most likely stayed at Hotel C.

Chapter Exercises

4.147 a. $S = \{$BL, BM, BH, GL, GM, GH, EL, EM, EH$\}$

b. $A = \{$EL, EM, EH$\}$, $B = \{$BH, GH, EH$\}$,
$C = \{$BL, BM, BH, GL, EL$\}$, $D = \{$GM$\}$

c. $A \cup B = \{$BH, EH, EL, EM, GH$\}$,
$B \cup C = \{$BH, BL, BM, EH, EL, GH, GL$\}$,
$D' = \{$BH, BL, BM, EH, EL, EM, GH, GL$\}$

d. $A \cap B = \{$EH$\}$, $C \cap D = \{\ \}$,
$(B \cup D)' = \{$BL, BM, EL, EM, GL$\}$

4.149 a. $P(A) = 0.167 + 0.057 + 0.010 = 0.234$
$P(B) = 1 - 0.080 = 0.920$
$P(C) = 0.193$

b. $P(A \cap B) = 0.167 + 0.057 + 0.010 = 0.234$
$P(A \cup C) = 0.234 + 0.193 = 0.427$
$P(B \cup C) = 1 - 0.08 = 0.92$

c. $P(C') = 1 - P(C) = 1 - 0.193 = 0.807$
$P(A' \cup B) = P(S) = 1.00$
$P(B' \cap C') = P(\text{Associate degree}) = 0.080$

4.151 Let B_i = person i owns savings bonds.

a. $P(B_1 \cap B_2 \cap B_3 \cap B_4) = P(B_1)\, P(B_2)\, P(B_3)\, P(B_4)$
$= (0.20)(0.20)(0.20)(0.20) = 0.0016$

b. $P(B_1' \cap B_2' \cap B_3' \cap B_4') = P(B_1')\, P(B_2')\, P(B_3')\, P(B_4')$
$= (0.80)(0.80)(0.80)(0.80) = 0.4096$

c. $P(2 \text{ of the } 4 \text{ own bonds})$
$= P(B_1 \cap B_2 \cap B_3' \cap B_4') + \cdots$
$= 6(0.20)^2(0.80)^2 = 0.1536$

4.153 a. $P(T \cap C) = 0.06$

b. $P(H \mid T') = 0.12/0.49 = 0.2449$

c. $P(T \mid G') = 0.46/0.77 = 0.5974$

d. $P(T_1) = P(T_2) = 0.51$
$P(T_1 \cap T_2') + P(T_1' \cap T_2)$
$= P(T_1)\, P(T_2') + P(T_1')\, P(T_2)$
$= (0.51)(0.49) + (0.49)(0.51) = 0.4998$

4.155 Let C_i = commuter i uses public transportation.

a. $P(C_1 \cap C_2 \cap C_3) = P(C_1)\, P(C_2)\, P(C_3)$
$= (0.546)(0.377)(0.327) = 0.0673$

b. $P(C_1' \cap C_2' \cap C_3') = P(C_1')\, P(C_2')\, P(C_3')$
$= (0.454)(0.623)(0.673) = 0.1904$

c. $P(C_1' \cap C_2 \cap C_3') = P(C_1')\, P(C_2)\, P(C_3')$
$= (0.454)(0.377)(0.673) = 0.1152$

d. $P(C_1 \cap C_2 \cap C_3') + P(C_1 \cap C_2' \cap C_3) + P(C_1' \cap C_2 \cap C_3)$
$= P(C_1)\, P(C_2)\, P(C_3') + P(C_1)\, P(C_2')\, P(C_3)$
$\quad + P(C_1')\, P(C_2)\, P(C_3)$
$= (0.546)(0.377)(0.673) + (0.546)(0.623)(0.327)$
$\quad + (0.454)(0.377)(0.327)$
$= 0.1385 + 0.1112 + 0.0560 = 0.3057$

4.157 Let H_i = person i received a heart from someone without health insurance.

a. $P(H_1 \cap H_2 \cap H_3 \cap H_4) = P(H_1)\, P(H_2)\, P(H_3)\, P(H_4)$
$= (0.25)(0.25)(0.25)(0.25) = 0.0039$

b. $P(H_1' \cap H_2' \cap H_3' \cap H_4') = P(H_1')\, P(H_2')\, P(H_3')\, P(H_4')$
$= (0.75)(0.75)(0.75)(0.75) = 0.3164$

c. $P(2 \text{ of } 4 \text{ received hearts from uninsured})$
$= P(H_1 \cap H_2 \cap H_3' \cap H_4') + \cdots$
$= P(H_1)\, P(H_2)\, P(H_3')\, P(H_4') + \cdots$
$= 6(0.25)^2(0.75)^2 = 0.2109$

4.159

a. $P(A \cap 0) = P(0)\, P(A \mid 0) = (0.80)(0.20) = 0.16$

b. $P(A) = P(0 \cap A) + P(1 \cap A) + P(2 \cap A)$
$= P(0)\, P(A \mid 0) + P(1)\, P(A \mid 1) + P(2)\, P(A \mid 2)$
$= (0.80)(0.20) + (0.15)(0.40) + (0.05)(0.90)$
$= 0.16 + 0.06 + 0.045 = 0.265$

c. $P(1 \mid A) = \frac{P(1 \cap A)}{P(A)} = \frac{0.06}{0.265} = 0.2264$

4.161 a. $P(\text{CC} \cap C') = P(\text{CC})\, P(C' \mid \text{CC})$
$= (0.07)(0.90) = 0.0630$

b. $P(C') = P(\text{Cash} \cap C') + P(\text{ATM} \cap C') + P(\text{CC} \cap C')$
$= P(\text{Cash})\, P(C' \mid \text{Cash}) + P(\text{ATM})\, P(C' \mid \text{ATM})$
$\quad + P(\text{CC})\, P(C' \mid \text{CC})$
$= (0.55)(0.25) + (0.38)(0.65) + (0.07)(0.90)$
$= 0.1375 + 0.2470 + 0.0630 = 0.4475$

c. $P(\text{ATM} \mid C') = \frac{P(\text{ATM} \cap C')}{P(C')} = \frac{0.2470}{0.4475} = 0.5520$

4.163 a.

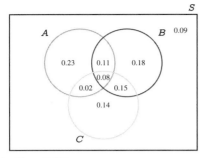

b. P(just A) = 0.23

c. $P(A' \cap B' \cap C') = 0.09$

d. $P(A \mid C) = \frac{P(A \cap C)}{P(C)} = \frac{0.10}{0.39} = 0.2564$

$P[B \mid (A \cap C)] = \frac{0.08}{0.10} = 0.80$

$P[(A \cap B \cap C) \mid A] = \frac{0.08}{0.44} = 0.1818$

4.165 a. $P(H) = 32484/638004 = 0.0509$

b. $P(W \mid 21\text{-}3) = 14486/25931 = 0.5586$

c. P(Unknown $\mid$ AA) $= 38765/243406 = 0.1593$

d. $P(22 \mid W') = 50676/315230 = 0.1608$

e. $P(H' \mid 24') = 604953/637409 = 0.9491$

f. $P(W) = 0.5059$, $P(TR) = 0.2275$
$P(W \cap TR) = 62348/638004 = 0.0977$
$P(W)\,P(TR) = (0.5059)(0.2275) = 0.1151$
$\neq P(W \cap TR)$
The events White and All other TR are not independent.

g. $P(21 \cap W) = 0.1892$
P(both stopped for Title 21 and White)
$= (0.1892)(0.1892) = 0.0358$

Chapter 5: Variables and Discrete Probability Distributions

Section 5.1: Random Variables

5.1 a. Discrete **b.** Continuous **c.** Continuous **d.** Discrete **e.** Discrete **f.** Continuous **g.** Discrete **h.** Discrete

5.3 a. Discrete **b.** Continuous **c.** Continuous **d.** Discrete **e.** Continuous **f.** Discrete

5.5 Y is continuous since it is a measurement of a length of time.

5.7 X is continuous since it is a measurement of distance.

Section 5.2: Probability Distributions for Discrete Random Variables

5.9 a. $p(5) = 1 - 0.93 = 0.07$

b. $P(2 \leq X \leq 6) = p(2) + p(3) + p(4) + p(5) + p(6)$
$= 0.20 + 0.15 + 0.12 + 0.07 + 0.08 = 0.62$

$P(2 < X \leq 6) = p(3) + p(4) + p(5) + p(6)$
$= 0.15 + 0.12 + 0.07 + 0.08 = 0.42$

c. $P(X < 4) = p(1) + p(2) + p(3)$
$= 0.35 + 0.20 + 0.15 = 0.70$

d. $P(X = 1, 7) = p(1) + p(7) = 0.35 + 0.03 = 0.38$

5.11 a. $P(X \geq 0) = p(0) + p(1) + p(2) + p(3)$
$= 0.30 + 0.05 + 0.10 + 0.20 = 0.65$

$P(X > 0) = p(1) + p(2) + p(3)$
$= 0.05 + 0.10 + 0.20 = 0.35$

b. $P(X^2 > 1) = p(-3) + p(-2) + p(2) + p(3)$
$= 0.20 + 0.10 + 0.10 + 0.20 = 0.60$

c. $P(X \geq 2 \,|\, X \geq 0) = \frac{P(X \geq 2 \cap X \geq 0)}{P(X \geq 0)}$

$= \frac{P(X \geq 2)}{P(X \geq 0)} = \frac{0.30}{0.65} = 0.4615$

d. Probability histogram:

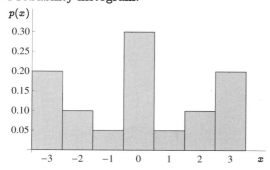

5.13 a. Probability distribution:

x	1	2	3	4
$p(x)$	0.01	0.08	0.27	0.64

b. Probability histogram:

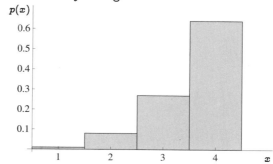

5.15 a. Probability distribution in tabular format:

x	1	2	3	4	5	6
$p(x)$	0.0179	0.0536	0.1071	0.1786	0.2679	0.3750

This is a valid probability distribution since $0 \leq p(x) \leq 1$ and $\sum p(x) = 1$.

b. $P(X = 4) = p(4) = 0.1786$

c. $P(X > 2) = p(3) + p(4) + p(5) + p(6)$
$= 0.1071 + 0.1786 + 0.2679 + 0.3750 = 0.9286$

d. $P(X = 3, 4) = p(3) + p(4)$
$= 0.1071 + 0.1786 = 0.2857$

e. Probability histogram:

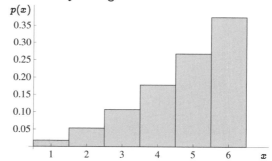

5.17

a. $P(Y > 0) = 1 = P(Y = 0) = 1 - 0.65 = 0.35$

b. $P(Y \leq 1000) = p(0) + p(500) + p(1000)$
$= 0.65 + 0.20 + 0.10 = 0.95$

c. $P(Y = 5000) = p(5000) = 0.04$

d. $P(Y_1 = 1000 \cap Y_2 = 1000)$
$= P(Y_1 = 1000) P(Y_2 = 1000) = (0.10)(0.10) = 0.01$

e. P(at least 1 driver paid 500 or more)
$= 1 - $ P(both drivers paid 0)
$= 1 - P(Y_1 = 0 \cap Y_2 = 0)$
$= 1 - P(Y_1 = 0) P(Y_2 = 0)$
$= 1 - (0.65)(0.65) = 1 - 0.4225 = 0.5775$

5.19 a.

x	0	1	2	3
$p(x)$	0.343	0.441	0.189	0.027

Probability histogram:

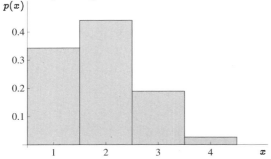

b. $P(X = 3) = p(3) = 0.027$

c. $P(X \geq 1) = 1 - P(X = 0) = 1 - 0.343 = 0.657$

5.21 $P(X = 0) = \dfrac{\binom{2}{0}\binom{4}{2}}{\binom{6}{2}} = \dfrac{6}{15} = 0.4000$

$P(X = 1) = \dfrac{\binom{2}{1}\binom{4}{1}}{\binom{6}{2}} = \dfrac{8}{15} = 0.5333$

$P(X = 2) = \dfrac{\binom{2}{2}\binom{4}{0}}{\binom{6}{2}} = \dfrac{1}{15} = 0.0667$

Probability distribution:

x	0	1	2
$p(x)$	0.4000	0.5333	0.0667

5.23 a. $P(X > 2) = p(3) + p(4) + p(5)$
$= 0.07 + 0.02 + 0.01 = 0.10$

b. $P(X \neq 2) = 1 - P(X = 2) = 1 - 0.35 = 0.65$

c. $P(X_1 = 1 \cap X_2 = 1) = P(X_1 = 1)\,P(X_2 = 1) =$
$(0.55)(0.55) = 0.3025$

d. P(total bags at least 8)
$= P(X_1 = 3 \cap X_2 = 5) + P(X_1 = 4 \cap X_2 = 4)$
$\quad + P(X_1 = 4 \cap X_2 = 5) + P(X_1 = 5 \cap X_2 = 3)$
$\quad + P(X_1 = 5 \cap X_2 = 4) + P(X_1 = 5 \cap X_2 = 5)$
$= 0.0007 + 0.0004 + 0.0002$
$\quad + 0.0007 + 0.0002 + 0.0001$
$= 0.0023$

e. Probability distribution:

y	50	100	150	200	250
$p(y)$	0.55	0.35	0.07	0.02	0.01

5.25 a.

Envelope 1	Envelope 2	Probability	m
100	100	0.0667	100
100	250	0.0667	250
100	500	0.0667	500
100	500	0.0667	500
100	1000	0.0667	1000
100	250	0.0667	250
100	500	0.0667	500
100	500	0.0667	500
100	1000	0.0667	1000
250	500	0.0667	500
250	500	0.0667	500
250	1000	0.0667	1000
500	500	0.0667	500
500	1000	0.0667	1000
500	1000	0.0667	1000

Probability distribution:

m	100	250	500	1000
$p(m)$	0.0667	0.1333	0.4667	0.3333

b. $P(M_1 = 1000 \cap M_2 = 1000)$
$= P(M_1 = 1000)\,P(M_2 = 1000)$
$= (0.3333)(0.3333) = 0.1111$

Section 5.3: Mean, Variance, and Standard Deviation for a Discrete Random Variable

5.27

$x^2 \cdot p(x)$	x^2	$p(x)$	x	$x \cdot p(x)$
0.40	4	0.10	2	0.20
2.56	16	0.16	4	0.64
7.20	36	0.20	6	1.20
15.36	64	0.24	8	1.92
18.00	100	0.18	10	1.80
17.28	144	0.12	12	1.44
60.80				7.20

$\mu = 7.20$
$\sigma^2 = 60.80 - (7.20)^2 = 8.96$
$\sigma = \sqrt{8.96} = 2.9933$

5.29 a. $\mu = (-20)(0.30) + \cdots + (20)(0.30) = 0$
$\sigma^2 = 270 - 0^2 = 270$
$\sigma = \sqrt{270} = 16.4317$

b. $P(\mu - 2\sigma \leq Y \leq \mu + 2\sigma)$
$= P(-32.8634 \leq Y \leq 32.8634)$
$= p(-20) + p(-10) + p(0) + p(10) + p(20) = 1$

c. $P(Y \geq \mu) = P(Y \geq 0)$
$= p(0) + p(10) + p(20) = 0.10 + 0.15 + 0.30 = 0.55$

$P(Y > \mu) = P(Y > 0)$
$= p(10) + p(20) = 0.15 + 0.30 = 0.45$

5.31 a. $\mu = 1(0.05) + \cdots + 21(0.05) = 7.35$
$\sigma^2 = 78.65 - (7.35)^2 = 24.6275$
$\sigma = \sqrt{24.6275} = 4.9626$

b. $P(X < \mu - \sigma) + P(X > \mu + \sigma)$
$= P(X < 2.3874) + P(X > 12.3126)$
$= p(1) + p(2) + p(13) + p(21)$
$= 0.05 + 0.10 + 0.20 + 0.05 = 0.40$

c. $P(X \le \mu + 2\sigma) = P(X \le 17.2752)$
$= 1 - P(X > 17.2752) = 1 - p(21) = 1 - 0.05 = 0.95$

5.33

a. $\mu = (0.25)(0.04) + \cdots + (2.50)(0.01) = 1.085$

b. $\sigma^2 = 1.3275 - (1.085)^2 = 0.1503$
$\sigma = \sqrt{0.1503} = 0.3877$

c. $P(\mu - \sigma \le X \le \mu + \sigma)$
$= P(0.6973 \le X \le 1.4727)$
$= p(1.00) + p(1.25) = 0.50 + 0.20 = 0.70$

d. $P(X < \mu + \sigma) = P(X < 1.4727)$
$= p(0.25) + p(0.50) + p(1.00) + p(1.25)$
$= 0.04 + 0.10 + 0.50 + 0.20 = 0.84$

5.35 a. $\mu = (10)(0.30) + \cdots + (30)(0.08) = 15.55$
$\sigma^2 = 282.25 - (15.55)^2 = 40.4475$
$\sigma = \sqrt{40.4475} = 6.3598$

b. $P(\mu - \sigma \le X \le \mu + \sigma) = P(9.1902 \le X \le 21.9098)$
$= p(10) + p(12) + p(15) + p(20)$
$= 0.30 + 0.25 + 0.15 + 0.12 = 0.82$

c. $P(\mu - 2\sigma \le X \le \mu + 2\sigma) = P(2.8304 \le X \le 28.2696)$
$= 1 - p(30) = 1 - 0.08 = 0.92$

d. A sunlamp lasts for $(100)(60) = 6000$ minutes. The mean tanning session lasts 15.55 minutes. A sunlamp lasts $6000/15.55 = 385.85$ sessions, or approximately 386.

5.37 a. $\mu = 1(0.002) + \cdots + 12(0.048) = \mu = 7.998$
$\sigma^2 = 70.674 - (7.998)^2 = 6.706$
$\sigma = \sqrt{6.706} = 2.5896$

b. $P(\mu - \sigma \le X \le \mu + \sigma) = P(5.4084 \le X \le 10.5876)$
$= p(6) + p(7) + p(8) + p(9) + p(10)$
$= 0.090 + 0.100 + 0.120 + 0.140 + 0.150 = 0.600$

c. $P(X < \mu - 2\sigma) + P(X > \mu + 2\sigma)$
$= P(X < 2.8188) + P(X > 13.1772)$
$= p(1) + p(2) = 0.002 + 0.010 = 0.012$

P(both more than 2 std dev away from the mean)
$= (0.012)(0.012) = 0.000144$

5.39 a. $\mu_X = 10(0.075) + \cdots + 16(0.040) = 12.725$
$\sigma_X^2 = 164.085 - (12.725)^2 = 2.1594$
$\sigma_X = \sqrt{2.1594} = 1.4695$

b. $P(X \ge \mu - \sigma) = P(X \ge 11.2555)$
$= 1 - (p(10 + p(11)) = 1 - (0.075 + 0.115)$
$= 1 - 0.190 = 0.810$

c. Probability distribution for Y:

y	11	13	15	17	19	21	23
$p(y)$	0.075	0.115	0.275	0.225	0.205	0.065	0.040

$\mu_Y = 11(0.075) + \cdots + 23(0.040) = 16.45$
$\sigma_Y^2 = 279.24 - (16.45)^2 = 8.6375$
$\sigma_Y = \sqrt{8.6375} = 2.939$

5.41 a. Probability distribution in tabular format:

x	-3	-2	-1	0
$p(x)$	0.2105	0.1579	0.1053	0.0526

x	1	2	3
$p(x)$	0.1053	0.1579	0.2105

This is a valid probability distribution since $0 \le p(x) \le 1$ and $\sum p(x) = 1$.

b. $\mu = (-3)(0.2105) + \cdots + (3)(0.2105) = 0$
$\sigma^2 = 5.2632 - 0^2 = 5.2632$
$\sigma = \sqrt{5.2632} = 2.2942$

c. $P(X_1 = 0 \cap X_2 = 0) = P(X_1 = 0) P(X_2 = 0)$
$= (0.0526)(0.0526) = 0.0028$

5.43 $\mu_y = a\mu_x + b$, $\sigma_Y^2 = a^2 \sigma_X^2$

5.45 $Y = \frac{X - \mu_X}{\sigma_X} = \frac{1}{\sigma_X} X - \frac{\mu_X}{\sigma_X}$

Using the results from Exercise 5.43:

$\mu_Y = \frac{1}{\sigma_X} \mu_X - \frac{\mu_X}{\sigma_X} = 0$

$\sigma_Y^2 = \left(\frac{1}{\sigma_X}\right)^2 \sigma_X^2 = 1$

$\sigma_Y = \sqrt{1} = 1$

Section 5.4: The Binomial Distribution

5.47

a. $P(X \ge 12) = 1 - P(X \le 11) = 1 - 0.9435 = 0.0565$

b. $P(X \ne 10) = 1 - P(X = 10) = 1 - 0.1171 = 0.8829$

c. $P(X \le 15) = 0.9997$

d. $P(2 < X \le 8) = P(X \le 8) - P(X \le 2)$
$= 0.5956 - 0.0036 = 0.5920$

5.49 a. $\mu = (25)(0.80) = 20$
$\sigma^2 = (25)(0.80)(0.20) = 4.0$, $\sigma = \sqrt{4} = 2$.

b. $P(\mu - \sigma \le X \le \mu + \sigma) = P(18 \le X \le 22)$
$= P(X \le 22) - P(X \le 17)$

$= 0.9018 - 0.1091 = 0.7927$

c. $P(X < \mu - 2\sigma) + P(X > \mu + 2\sigma)$
$= P(X < 16) + P(X > 24)$
$= P(X \le 15) + (1 - P(X \le 24))$
$= 0.0173 + (1 - 0.9962) = 0.0211$

5.51 a.

x	$p(x)$	x	$p(x)$	x	$p(x)$
0	0.0010	1	0.0098	2	0.0439
3	0.1172	4	0.2051	5	0.2461
6	0.2051	7	0.1172	8	0.0439
9	0.0098	10	0.0010		

b. $\mu = 0(0.0010) + \cdots + 10(0.0010) = 5.0$
$\sigma^2 = 27.5 - (5.0)^2 = 2.5$
$\sigma = \sqrt{2.5} = \sigma = 1.5811$

c. $\mu = (10)(0.50) = 5$
$\sigma^2 = (10)(0.5)(0.5) = 2.5, \quad \sigma = \sqrt{2.5} = 1.5811$

5.53 Let X = the number of cars that start.
$X \sim B(15, 0.75)$
a. $P(X < 10) = P(X \le 9) = 0.1484$
b. $P(X > 12) = 1 - P(X \le 12) = 1 - 0.7639 = 0.2361$
c. Claim: $p = 0.75 \Rightarrow X \sim B(15, 0.75)$
Experiment: $x = 9$
Likelihood: $P(X \le 9) = 0.1484$
Conclusion: There is no evidence to suggest the claim is false.

5.55 Let X = the number of damaged cars.
$X \sim B(20, 0.60)$
a. $P(X = 10) = 0.1171$
b. $P(X \ge 15) = 1 - P(X \le 14) = 1 - 0.8744 = 0.1256$
c. $P(X \le 12) = 0.5841$
d. Claim: $p = 0.60 \Rightarrow X \sim B(20, 0.60)$
Experiment: $x = 19$
Likelihood: $P(X \ge 19) = 0.0005$
Conclusion: There is evidence to suggest the claim is false.

5.57 Let X = the number of commercial buildings with at least one violation. $X \sim B(30, 0.25)$
a. $\mu = (30)(0.25) = 7.5$
$\sigma^2 = (30)(0.25)(0.75) = 5.625, \quad \sigma = \sqrt{5.625} = 2.3717$
b. $P(\mu - \sigma \le X \le \mu + \sigma) = P(5.1283 \le X \le 9.8717)$
$= P(6 \le X \le 9) = P(X \le 9) - P(X \le 5)$
$= 0.8034 - 0.2026 = 0.6008$
c. $P(X < \mu - 2\sigma) + P(X > \mu + 2\sigma)$
$= P(X < 2.7566) + P(X > 12.2434)$
$= P(X \le 2) + (1 - P(X \le 12))$

$= 0.0106 + (1 - 0.9784) = 0.0106 + 0.0216 = 0.0322$

d. Claim: $p = 0.25 \Rightarrow X \sim B(30, 0.25)$
Experiment: $x = 10$
Likelihood: $P(X \ge 10) = 0.1966$
Conclusion: There is no evidence to suggest the claim is false.

5.59 Let X = the number of defective CDs.
$X \sim B(100, 0.02)$
a. $P(X = 0) = 0.1326$
b. $\mu = (100)(0.02) = 2$
c. $P(X > 2) = 1 - P(X \le 2) = 1 - 0.6767 = 0.3233$
d. Claim: $p = 0.02 \Rightarrow X \sim B(100, 0.02)$
Experiment: $x = 6$
Likelihood: $P(X \ge 6) = 0.0155$
Conclusion: There is evidence to suggest the claim is false.

5.61 Let X = the number of people who get colds.
$X \sim B(30, 0.2279)$
a. $\mu = (30)(0.2279) = 6.837$
$\sigma^2 = (30)(0.2279)(0.7721) = 5.2788,$
$\sigma = \sqrt{5.2788} = 2.2976$
b. $P(X > 10) = 1 - P(X \le 10) = 1 - 0.9393 = 0.0607$
c. $P(X = 14) = 0.0024$
d. Claim: $p = 0.2279 \Rightarrow X \sim B(30, 0.2279)$
Experiment: $x = 5$
Likelihood: $P(X \le 5) = 0.29$
Conclusion: There is no evidence to suggest the claim is false.

5.63 Let X = the number of Americans who are pessimistic about the economy. $X \sim B(50, 0.44)$
a. $P(X = 30) = 0.0087$
b. $P(X > 23) = 1 - P(X \le 23) = 1 - 0.6669 = 0.3331$
c. Claim: $p = 0.44 \Rightarrow X \sim B(50, 0.44)$
Experiment: $x = 17$
Likelihood: $P(X \le 17) = 0.0990$
Conclusion: There is no evidence to suggest the claim is false.

5.65 Let X = the number of Blu-ray players purchased. $X \sim B(20, 0.10)$
a. $P(X \ge 5) = 1 - P(X \le 4) = 1 - 0.9568 = 0.0432$
b. $\mu = (20)(0.10) = 2$
$P(X = 2) = 0.2852$
c. $P(X = 0 \mid X \le 4) = \frac{P(X=0 \cap X \le 4)}{P(X \le 4)} = \frac{P(X=0)}{P(X \le 4)}$
$= \frac{0.1216}{0.9568} = 0.1271$

5.67 Let X = the number of machines that detect the explosive.   **a.** $X \sim B(3, 0.60)$

$P(X = 1) = 0.2880$

$P(X = 0) = 0.0640$

b. $Y \sim B(4, 0.60)$

$P(Y \geq 1) = 1 - P(Y = 0) = 1 - 0.0256 = 0.9744$

c. $W \sim B(5, 0.60)$

$P(W \geq 1) = 1 - P(W = 0) = 1 - 0.0102 = 0.9898$

d. $V \sim B(n, 0.60)$

$P(V \geq 1) = 1 - P(V = 0) \geq 0.999$

$\Rightarrow P(V = 0) \leq 0.001$

$\Rightarrow (0.40)^n \leq 0.001$

$\Rightarrow n(\ln 0.40) \leq \ln 0.001$

$\Rightarrow n \geq 7.53 \Rightarrow n \geq 8$

5.69 Let X = the number of defective switches.

a. $X \sim B(25, 0.02)$

$P(X \geq 4) = 1 - P(X \leq 3) = 1 - 0.9986 = 0.0014$

b. $X \sim B(25, 0.05)$

$P(X \leq 3) = 0.9659$

c. $X \sim B(25, 0.07)$

$P(X \leq 3) = 0.9064$

Section 5.5: Other Discrete Distributions

5.71 a. $P(X = 4) = 0.0961$

b. $P(X \geq 3) = 1 - P(X \leq 2) = 1 - 0.5775 = 0.4225$

c. $P(X \leq 2) = 0.5775$

d. $\mu = 1/0.35 = 2.8571$

$P(X \geq 2.8571) = 1 - P(X \leq 2)$

$= 1 - 0.5775 = 0.4225$

5.73 a. $P(X = 0) = 0.1353$

b. $P(2 \leq X \leq 8) = P(X \leq 8) - P(X \leq 1)$

$= 0.9998 - 0.4060 = 0.5938$

c. $P(X > 5) = 1 - P(X \leq 5) = 1 - 0.9834 = 0.0166$

d. $P(X \leq 6) = 0.9955$

5.75 a. $P(X = 2) = \frac{\binom{6}{2}\binom{6}{3}}{\binom{12}{5}} = 0.3788$

b. $P(X = 5) = \frac{\binom{6}{5}\binom{6}{0}}{\binom{12}{5}} = 0.0076$

c. $\mu = 5\frac{6}{12} = 2.5$

$\sigma^2 = \frac{7}{11} 5 \frac{6}{12} \left(1 - \frac{6}{12}\right) = 0.7955$

$\sigma = \sqrt{0.7955} = 0.8919$

5.77 Let X = the number of times a person must take the test before passing. X is a geometric random variable with $p = 0.70$.

a. $P(X = 3) = 0.0630$

b. $P(X \geq 2) = 1 - P(X \leq 1) = 1 - 0.70 = 0.30$

c. $\mu = 1/p = 1/0.7 = 1.4286$

d. Let $Y = 50X$.

$E(Y) = 50E(X) = 50(1.4286) = 71.4286$

5.79 Let X = the number of children examined until a heart murmur is detected. X is a geometric random variable with $p = 0.90$.

a. $P(X = 2) = 0.09$

b. $P(X = 7) = 0.0000009$

c. $\mu = 1/p = 1/0.90 = 1.1111$

d. $P(X \geq 5) = 1 - P(X \leq 4) = 1 - 0.9999 = 0.0001$

5.81 Let X = the number of injuries per 200,000 employee hours. X is a Poisson random variable with $\lambda = 1.4$.

a. $P(X \leq 4) = 0.9857$

b. $P(X > \mu + 2\sigma) = P(X > 3.7664)$

$= 1 - P(X \leq 3) = 1 - 0.9463 = 0.0537$

c. $P(5\ 200{,}000\text{-hour periods without injuries})$

$= [P(X = 0)]^5 = (0.2466)^5 = 0.0009119$

5.83 Let X = the number of dead pixels in an LCD monitor. X is a Poisson random variable with $\lambda = 2.5$.

a. $P(X = 0) = 0.0821$

b. $P(X > \mu + 3\sigma) = P(X > 7.2434)$

$= 1 - P(X \leq 7) = 1 - 0.9958 = 0.0042$

c. $P(\mu - 2\sigma \leq X \leq \mu + 2\sigma)$

$= P(-0.6623 \leq X \leq 5.6623)$

$= P(X \leq 5) = 0.9580$

5.85 Let X = the number of lobstermen fined. X is a hypergeometric random variable with $n = 4$, $N = 15$, and $M = 5$.

a. $P(X = 2) = \frac{\binom{5}{2}\binom{10}{2}}{\binom{15}{4}} = 0.3297$

b. $P(X = 4) = \frac{\binom{5}{4}\binom{10}{0}}{\binom{15}{4}} = 0.0037$

c. $P(X \geq 1) = 1 - P(X = 0)$

$= 1 - \frac{\binom{5}{0}\binom{10}{4}}{\binom{15}{4}} = 1 - 0.1538 = 0.8462$

5.87 Let X = the number of winning bottles. X is a hypergeometric random variable with $n = 6$, $N = 20$, and $M = 2$.

a. $P(X = 0) = \frac{\binom{2}{0}\binom{18}{6}}{\binom{20}{6}} = 0.4789$

b. $P(X = 2) = \frac{\binom{2}{2}\binom{18}{4}}{\binom{20}{6}} = 0.0789$

c. $\mu = 6\frac{2}{20} = 0.60$

d. $\mu = 1 = n\frac{2}{20} \Rightarrow n = 10$

5.89 Let X = the number of telephone orders until the first error. X is a geometric random variable with $p = 0.20$.

a. $P(X = 3) = 0.1280$

b. $P(2 \leq X \leq 6) = P(X \leq 6) - P(X \leq 1)$
$= 0.7379 - 0.2000 = 0.5379$

c. $P(X \geq 7) = 1 - P(X \leq 6) = 1 - 0.7379 = 0.2621$

d. $P(X \geq 4) = 1 - P(X \leq 3) = 1 - 0.4880 = 0.5120$

e. $P(X \geq 8 \mid X \geq 5) = \frac{P(X \geq 8 \cap X \geq 5)}{P(X \geq 5)} = \frac{P(X \geq 8)}{P(X \geq 5)}$
$= \frac{1 - P(X \leq 7)}{1 - P(X \leq 4)} = \frac{0.2097}{0.4096} = 0.5120$

5.91 Let X = the number of times a person washes their primary vehicle per year. X is a Poisson random variable with $\lambda = 4$.

a. $P(X = 0) = 0.0183$, $P(X = 1) = 0.0733$, $P(X = 2) = 0.1465$

b. Let Y = the total number of washes per year.

$P(Y = 0) = P(X_1 = 0 \cap X_2 = 0)$
$= (0.0183)(0.0183) = 0.0003$

$P(Y = 1)$
$= P(X_1 = 0 \cap X_2 = 1) + P(X_1 = 1 \cap X_2 = 0)$
$= 2(0.0183)(0.0733) = 0.0027$

$P(Y = 2) = P(X_1 = 0 \cap X_2 = 2)$
$\quad + P(X_1 = 1 \cap X_2 = 1) + P(X_1 = 2 \cap X_2 = 0)$
$= 0.0027 + 0.0054 + 0.0027 = 0.0107$

c. Let W = the number of times a person washes their primary vehicle in a two-year period. W is a Poisson random variable with $\lambda = 8$.

$P(W = 0) = 0.0003$, $P(W = 1) = 0.0027$, $P(W = 2) = 0.0107$

These probabilities are the same. The Poisson random variable is additive.

Chapter Exercises

5.93 Let X = the number of automobiles that pass the enhanced emissions test. $X \sim B(30, 0.93)$

a. $P(X = 28) = 0.2794$

b. $P(X \geq 25) = 1 - P(X \leq 24) = 1 - 0.0162 = 0.9838$

c. Claim: $p = 0.93 \Rightarrow X \sim B(30, 0.93)$
Experiment: $x = 20$
Likelihood: $P(X \leq 20) = 0.00002296$
Conclusion: There is evidence to suggest the claim is false.

5.95 a. $\mu = 1(0.02) + \cdots + 10(0.02) = 5.93$
$\sigma^2 = 39.63 - (5.93)^2 = 4.4651$
$\sigma = \sqrt{4.4651} = 2.1131$

b. $P(X > 7) = p(8) + p(9) + p(10) = 0.24$

c. $P(X < \mu - \sigma) = P(X < 3.8169)$
$= p(1) + p(2) + p(3) = 0.16$

d. Let F_i = person i carries fewer than 6 cards.
$P(F_i) = P(X < 6) = p(1) + \cdots + p(5) = 0.37$
$P(\text{at least 1 carries fewer than 6})$
$= 1 - P(0 \text{ carry fewer than 6})$
$= 1 - P(F_1' \cap F_2' \cap F_3')$
$= 1 - (0.63)(0.63)(0.63) = 0.75$

5.97 Let X = the number of missing pieces per packaged grill. X is a Poisson random variable with $\lambda = 0.7$.

a. $P(X = 0) = 0.4966$

b. $P(X > 5) = 1 - P(X \leq 5) = 1 - 0.99991 = 0.00009$

c. $P(X_1 \leq 1 \cap X_2 \leq 1 \cap X_3 \leq 1)$
$= P(X_1 \leq 1) P(X_2 \leq 1) P(X_3 \leq 1)$
$= (0.8442)^3 = 0.6016$

5.99 Let X = the number of rescues per hour. X is a Poisson random variable with $\lambda = 1.8$.

a. $P(X = 0) = 0.1653$

b. $P(2 \leq X \leq 6) = P(X \leq 6) - P(X \leq 1)$
$= 0.9974 - 0.4628 = 0.5346$

c. Claim: $\mu = 1.8 \Rightarrow X$ is a Poisson random variable with $\lambda = 1.8$.
Experiment: $x = 9$
Likelihood: $P(X \geq 9) = 0.0001$
Conclusion: There is evidence to suggest the claim is false. There is some justification for hiring more lifeguards.

d. Let Y = the number of rescues per 10,000 swimmers. Y is a Poisson random variable with $\lambda = 9.1$.

$P(Y = 0) = 0.0001167$

5.101 Let X = the number of new business that survive their first year. $X \sim B(35, 0.80)$

a. $P(X \geq 30) = 1 - P(X \leq 29)$
$= 1 - 0.7279 = 0.2721$

b. $P(X = 35) = 0.0004$

c. Expected number to survive $= (35)(0.80) = 28$
Expected number to close $= 35 - 28 = 7$
Expected number to lose their jobs $= 7(12) = 84$

5.103 Let X = the number of people who describe the flight as a nightmare. $X \sim B(50, 0.84)$

a. $\mu = (50)(0.84) = 42$

$\sigma^2 = (50)(0.84)(0.16) = 6.72$, $\sigma = \sqrt{6.72} = 2.5923$

b. $P(X \geq 40) = 1 - P(X \leq 39)$

$= 1 - 0.1661 = 0.8339$

c. $P(\mu - 2\sigma \leq X \leq \mu + 2\sigma) = P(36.8 \leq X \leq 47.2)$

$= P(X \leq 47) - P(X \leq 36)$

$= 0.9910 - 0.0223 = 0.9687$

5.105 Let $X =$ the number of adults who think PACs have too much power. $X \sim B(50, 0.85)$

a. $P(X \geq 45) = 1 - P(X \leq 44) = 1 - 0.7806 = 0.2194$

b. $P(38 \leq X \leq 42) = P(X \leq 42) - P(X \leq 37)$

$= 0.4812 - 0.0301 = 0.4512$

c. Claim: $p = 0.85 \Rightarrow X \sim B(50, 0.85)$

Experiment: $x = 35$

Likelihood: $P(X \leq 35) = 0.0053$

Conclusion: There is evidence to suggest the claim is false, that the poll results are wrong.

5.107 a. $\mu = 1(1/5) + \cdots + 5(1/5) = 3$

$\sigma^2 = 11 - 3^2 = 2$

$\sigma = \sqrt{2} = 1.4142$

b. $\mu = 1(1/6) + \cdots + 6(1/6) = 3.5$

$\sigma^2 = 15.1667 - (3.5)^2 = 2.9167$

$\sigma = \sqrt{2.9167} = 1.7078$

c. $\mu = 1(1/n) + \cdots + n(1/n) = (n+1)/2$

$\sigma^2 = \frac{(n+1)(2n+1)}{6} - \left(\frac{(n+1)}{2}\right)^2 = \frac{n^2-1}{12}$

$\sigma = \sqrt{(n^2 - 1)/12}$

5.109 a.

Tickets	x	Probability	Tickets	x	Probability
10 10	10	0.2500	10 20	20	0.1750
10 30	30	0.0500	10 40	40	0.0250
20 10	20	0.1750	20 20	20	0.1225
20 30	30	0.0350	20 40	40	0.0175
30 10	30	0.0500	30 20	30	0.0350
30 30	30	0.0100	30 40	40	0.0050
40 10	40	0.0250	40 20	40	0.0175
40 30	40	0.0050	40 40	40	0.0025

Probability distribution for X:

x	10	20	30	40
$p(x)$	0.2500	0.4725	0.1800	0.0975

b. $\mu = 10(0.2500) + \cdots + 40(0.0975) = 21.25$

$\sigma^2 = 532 - (21.25)^2 = 80.4375$

$\sigma = \sqrt{80.4375} = 8.9687$

c. $P(X \geq 20) = 1 - P(X = 10) = 1 - 0.25 = 0.75$

5.111 a. Answers will vary.

Complete shipments	Relative frequency	$p(y)$
0	0.0000	0.0000
1	0.0000	0.0007
2	0.0060	0.0052
3	0.1600	0.0234
4	0.0790	0.0701
5	0.1300	0.1471
6	0.2020	0.2207
7	0.2640	0.2365
8	0.1850	0.1774
9	0.0830	0.0887
10	0.0320	0.0266
11	0.0030	0.0036

b.

Complete shipments	Relative frequency	$p(y)$
0	0.0000	0.0000
1	0.0000	0.0000
2	0.0000	0.0001
3	0.0000	0.0008
4	0.0050	0.0040
5	0.0140	0.0142
6	0.0390	0.0392
7	0.0960	0.0840
8	0.1210	0.1417
9	0.1980	0.1889
10	0.1910	0.1983
11	0.1590	0.1623
12	0.1050	0.1014
13	0.0580	0.0468
14	0.0120	0.0150
15	0.0020	0.0030
16	0.0000	0.0003

c. $Y \sim B(25, 0.60)$

$P(Y = 15) = 0.1612$

$P(Y \leq 12) = 0.1538$

$P(Y > 16) = 1 - P(Y \leq 16) = 1 - 0.7265 = 0.2735$

$\mu = (25)(0.60) = 15$

Chapter 6: Continuous Probability Distributions

Section 6.1: Probability Distributions for Continuous Random Variables

6.1 a. Probability density function:

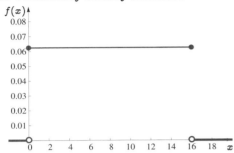

b. $\mu = \frac{0+16}{2} = 8$

$\sigma^2 = \frac{(16-0)^2}{12} = 21.3333$

$\sigma = \sqrt{21.3333} = 4.6188$

c. $P(X \geq 4) = (12)(1/16) = 0.75$

d. $P(2 \leq X < 12) = (10)(1/16) = 0.625$

e. $P(X \leq 7) = 7(1/16) = 0.4375$

6.3 a. $\mu = \frac{50+100}{2} = 75$

$\sigma^2 = \frac{(100-50)^2}{12} = 208.3333$

$\sigma = \sqrt{208.3333} = 14.4338$

b. $P(\mu - \sigma \leq X \leq \mu + \sigma)$
$= P(60.5662 \leq X \leq 89.4338)$
$= (28.8675)(0.02) = 0.5774$

c. $P(X \geq \mu + 2\sigma) = P(X \geq 103.8675) = 0$

d. $P(X \leq c) = (c-50)(0.02) = 0.20$
$\Rightarrow (c - 50) = 10 \Rightarrow c = 60$

6.5 a. $P(X \leq 1) = \left(\frac{1}{2}\right)(1)\left(\frac{1}{8}\right) = 0.0625$

b. $P(X > 3) = \left(\frac{1}{2}\right)(1)\left(\frac{3}{8} + \frac{4}{8}\right) = 0.4375$

c. $P(X > 4) = 0$

d. $P(2 \leq X \leq 3) = \left(\frac{1}{2}\right)(1)\left(\frac{2}{8} + \frac{3}{8}\right) = 0.3125$

e. $P(X \leq 2 \mid X \leq 3) = \frac{P(X \leq 2 \cap X \leq 3)}{P(X \leq 3)} = \frac{P(X \leq 2)}{P(X \leq 3)}$

$= \frac{\frac{1}{2} \cdot 2 \cdot \frac{2}{8}}{\frac{1}{2} \cdot 3 \cdot \frac{3}{8}} = 0.4444$

f. $P(X \leq c) = \left(\frac{1}{2}\right)(c)\left(\frac{c}{8}\right) = 0.5$

$\Rightarrow \frac{c^2}{8} = 1 \Rightarrow c^2 = 8$

$\Rightarrow c = \sqrt{8} = 2.8284$

The distribution is not symmetric. Therefore, the median is not $(0 + 4)/2 = 2$.

6.7 a. $P(W > 5.14) = (0.015)(14.2857) = 0.2143$

b. $P(C < 9.03) = (0.03)(10) = 0.3$

c. $P(5.11 \leq W \leq 5.13) \cdot P(9.04 \leq W \leq 9.06)$
$= (0.02)(14.2857) \cdot (0.02)(10) = 0.0571$

6.9 Let $X = $ the epoxy paint drying time and $h = 1/(60 - 30)$.

a. $P(X \leq 45) = (15)h = 0.5$

b. $P(40 \leq X \leq 50) = (10)h = 0.3333$

c. $P(X \geq t) = (60 - t)h = 0.75$
$\Rightarrow 60 - t = 22.5 \Rightarrow t = 37.5$

d. $P(X > 55) = (5)h = 0.1667$

6.11 a. $f(x) \geq 0$ and total area is $\frac{1}{2}(20)(0.10) = 1$.

b. $P(X \leq 5) = \left(\frac{1}{2}\right)(5)(f(0) + f(5))$
$= (0.50)(5)(0.100 + 0.075) = 0.4375$

c. $P(5 \leq X \leq 10) = \left(\frac{1}{2}\right)(5)(f(5) + f(10))$
$= (0.50)(5)(0.075 + 0.050) = 0.3125$

d. $P(X \leq t) = \left(\left(\frac{1}{2}\right)(t)(f(0) + f(t))\right) = 0.50$
$\Rightarrow (t)(0.100 - 0.005t + 0.100) = 1$
$\Rightarrow -0.005t^2 + 0.200t - 1 = 0 \Rightarrow t = 5.8579$

e. $p = P(X < 15)$
$= \left(\frac{1}{2}\right)(15)(f(0) + f(15)) = 0.9375$
Let $Y = $ the number of people who fall asleep within 15 minutes. $Y \sim B(20, 0.9375)$
$P(Y = 15) = 0.0056$
$P(Y \geq 18) = 1 - P(Y \leq 17) = 1 - 0.1259 = 0.8741$
$P(Y = 20) = 0.2751$

6.13 Let $X = $ number of pounds of candy purchased by the next customer.

a. $f(x) \geq 0$ and total area is
$(1)(0.25) + (1)(0.50) + (1)(0.25) = 1.00$.

b. $P(X \leq 2) = (1)(0.25) + (1)(0.50) = 0.75$

c. $P(X > 1) = 1 - (1)(0.25) = 0.75$

d. $P(X \leq 0.5 \mid X \leq 1.5) = \frac{P(X \leq 0.5)}{P(X \leq 1.5)}$
$= \frac{(0.5)(0.25)}{(1)(0.25) + (0.5)(0.50)} = \frac{0.125}{0.500} = 0.25$

6.15 a. $P(X \leq 20) = 1 - e^{-0.05 \cdot 20} = 0.6321$

b. $P(X > 30) = 1 - P(X \leq 30)$
$= 1 - (1 - e^{-0.05 \cdot 30}) = 1 - 0.7769 = 0.2231$

c. $P(15 \leq X \leq 30) = P(X \leq 30) - P(X \leq 15)$
$= (1 - e^{-0.05 \cdot 30}) - (1 - e^{-0.05 \cdot 15})$
$= 0.7769 - 0.5276 = 0.2492$

6.17 a. $P(X \leq 4) = 1 - e^{-4^2/8} = 0.8647$

b. $P(X > 2) = 1 - P(X \leq 2)$
$= 1 - (1 - e^{-2^2/8}) = 1 - 0.3935 = 0.6065$

c. $P(1 \leq X \leq 3) = P(X \leq 3) - P(X \leq 1)$
$= (1 - e^{-3^2/8}) - (1 - e^{-1^2/8})$
$= 0.6753 - 0.1175 = 0.5578$

d. $P(X \leq 2 \mid X \leq 4) = \frac{P(X \leq 2)}{P(X \leq 4)}$

$= \frac{1 - e^{-2^2/8}}{1 - e^{-4^2/8}} = \frac{0.3935}{0.8647} = 0.4551$

Section 6.2: The Normal Distribution

6.19

a.

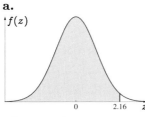

$P(Z \leq 2.16) = 0.9846$

b.

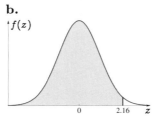

$P(Z < 2.16) = 0.9846$

c.

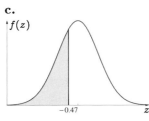

$P(Z \leq -0.47) = 0.3192$

d.

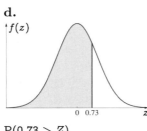

$P(0.73 > Z)$
$= P(Z < 0.73) = 0.7673$

e.

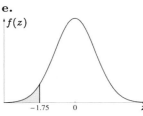

$P(-1.75 \geq Z)$
$= P(Z \leq -1.75) = 0.0401$

f.

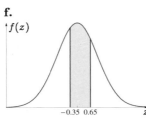
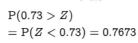

$P(-0.35 \leq Z \leq 0.65)$
$= P(Z \leq 0.65)$
$\quad - P(Z \leq -0.35)$
$= 0.7422 - 0.3632 = 0.3790$

g.

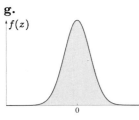

$P(Z < 5) \approx 1$

h.

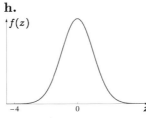

$P(Z \leq -4) \approx 0$

i.

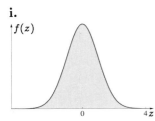

$P(Z \leq 4) \approx 1$

6.21 a. $P(-1.00 \leq Z \leq 1.00)$
$= P(Z \leq 1.00) - P(Z \leq -1.00)$
$= 0.8413 - 0.1587 = 0.6826$

b. $P(-2.00 \leq Z \leq 2.00)$
$= P(Z \leq 2.00) - P(Z \leq -2.00)$
$= 0.9772 - 0.0228 = 0.9544$

c. $P(-3.00 \leq Z \leq 3.00)$
$= P(Z \leq 3.00) - P(Z \leq -3.00)$
$= 0.9987 - 0.0013 = 0.9974$

These are the Empirical Rule probabilities.

6.23

a.

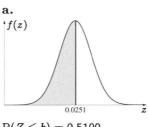

$P(Z \leq b) = 0.5100$
$b = 0.0251$

b.

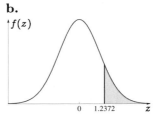

$P(Z > b) = 0.1080$
$P(Z \leq b) = 0.8920$
$b = 1.2372$

c.

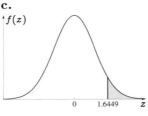

$P(Z \geq b) = 0.0500$
$P(Z \leq b) = 0.9500$
$b = 1.6449$

d.

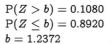

$P(Z \leq b) = 0.0100$
$b = -2.3263$

e.

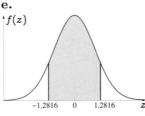

$P(-b \leq Z \leq b) = 0.8000$
$P(Z \leq -b) = 0.1000$
$b = 1.2816$

f.

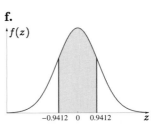

$P(-b < Z < b) = 0.6535$
$P(Z \leq -b) = 0.1733$
$b = 0.9412$

6.25 a. $P(Z \leq Q_1) = 0.25$
$\Rightarrow Q_1 = -0.6745, \; Q_3 = 0.6745$

b. $IQR = 0.6745 - (-0.6745) = 1.349$
$IF_L = -0.6745 - (1.5)(1.349) = -2.698$
$IF_H = 0.6745 + (1.5)(1.349) = 2.698$

c. $2P(Z < -2.698) = 2(0.0035) = 0.0070$

d. $OF_L = -0.6745 - (3)(1.349) = -4.7215$
$OF_H = -0.6745 + (3)(1.349) = 4.7215$

e. $2P(Z < -4.7215) = 0.00000234$

6.27

a.

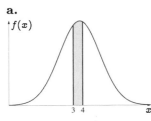

b.

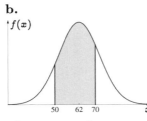

$P(3.0 \leq X \leq 4.0)$
$= 0.5559 - 0.3714 = 0.1845$

$P(50 < X < 70)$
$= 0.7881 - 0.1151 = 0.6730$

c.

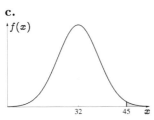

d.

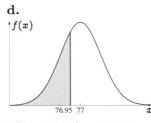

$P(X \geq 45)$
$= 1 - 0.9912 = 0.0088$

$P(X < 76.95) = 0.3085$

e.

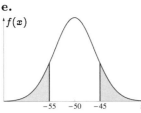

f.

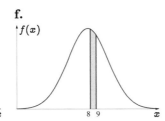

$P(X < -55 \cup X > -45)$
$= 0.1056 + 0.1056 = 0.2112$

$P(8 \leq X \leq 9)$
$= 0.6569 - 0.5460 = 0.1109$

6.29 a. $P(Z \leq Q_1) = 0.25$
$\Rightarrow \; Q_1 = 20.9531, \; Q_3 = 29.0469$

b. $IQR = 29.0469 - 20.9531 = 8.0938$
$IF_L = 20.9531 - (1.5)(8.0938) = 8.8124$
$IF_H = 29.0469 + (1.5)(8.0938) = 41.1876$

c. $2P(X < 8.8124) = 2(0.0035) = 0.0070$

d. $OF_L = 20.9531 - (3)(8.0938) = -3.3283$
$OF_H = 29.0469 + (3)(8.0938) = 53.3283$

e. $2P(X < -3.3283) = 0.00000234$

6.31 Let $X =$ the cost of a wedding.
$X \sim N(28732, 1500^2)$

a. $P(X > 32000) = P(Z > 2.18)$
$= 1 - P(Z \leq 2.18) = 1 - 0.9854 = 0.0146$

b. $P(25000 \leq X \leq 30000) = P(-2.49 \leq Z \leq 0.85)$
$= P(Z \leq 0.85) - P(Z \leq -2.49)$
$= 0.8023 - 0.0064 = 0.7959$

c. $P(X < 22000) = P(Z < -4.49) = 0.00000356$

6.33 Let $X =$ the head circumference of all men in the military. $X \sim N(22.05, 0.68^2)$

a. $P(X < 21) = P(Z < -1.54) = 0.0618$

b. $P(22 \leq X \leq 23) = P(-0.07 \leq Z \leq 1.40)$
$= 0.9192 - 0.4721 = 0.4471$

c. $P(X > 24) = P(Z > 2.87)$
$= 1 - P(Z \leq 2.87)$
$= 1 - 0.9979 = 0.0021$

d. $P(22.05 - c \leq X \leq 22.05 + c) = 0.95$
$\Rightarrow \; P(Z \leq -c/0.68) = 0.025$
$\Rightarrow \; -c/0.68 = -1.96 \; \Rightarrow \; c = 1.3328$
Interval: $(20.7172, 23.3828)$

6.35 Let $X =$ the salinity in a region of the tropical Pacific Ocean. $X \sim N(34.95, 0.52^2)$

a. $P(X > 36) = P(Z > 2.02)$
$= 1 - P(Z \leq 2.02)$
$= 1 - 0.9783 = 0.0217$

b. $P(X < 33.5) = P(Z < -2.79) = 0.0026$

c. $P(33 \leq X \leq 35) = P(-3.75 \leq Z \leq 0.10)$
$= P(Z \leq 0.01) - P(Z \leq -3.75)$
$= 0.5398 - 0.0001 = 0.5397$

d. $P(34.95 - s \leq X \leq 34.95 + s) = 0.50$
$\Rightarrow \; P(Z \leq -s/0.52) = 0.25$
$\Rightarrow \; -s/0.52 = -0.6745 \; \Rightarrow \; s = 0.3507$
Interval: $(34.5993, 35.3007)$. The endpoints of this interval are the quartiles for this distribution.

6.37 Let $X =$ the amount of selenium consumed from food per day for a woman over age 20.
$X \sim N(92.6, 18.4^2)$

a. $P(X \leq 100) = P(Z \leq 0.40) = 0.6554$

b. $P(60 \leq X \leq 75) = P(-1.77 \leq Z \leq -0.96)$
$= P(Z \leq -0.96) - P(Z \leq -1.77)$
$= 0.1685 - 0.0384 = 0.1301$

c. $P(X \geq 90) = P(Z \geq -0.14)$
$= 1 - P(Z \leq -0.14)$
$= 1 - 0.4443 = 0.5557$

d. $P(X \geq g) = 0.75 \; \Rightarrow \; P\left(Z \leq \frac{g-92.6}{18.4}\right) = 0.25$
$\Rightarrow \; \frac{g-92.6}{18.4} = -0.6745 \; \Rightarrow \; g = 80.1892$

6.39 Let $X =$ the time spent aloft in a hot-air balloon. $X \sim N(1.5, 0.45^2)$

a. $P(1 \leq X \leq 2) = P(-1.11 \leq Z \leq 1.11)$
$= P(Z \leq 1.11) - P(Z \leq -1.11)$
$= 0.8665 - 0.1335 = 0.7330$

b. $P(X > 1.25) = P(Z > -0.56)$
$= 1 - P(Z \leq -0.56)$
$= 1 - 0.2877 = 0.7123$

c. $P(X < t) = 0.10 \Rightarrow P\left(Z > \frac{t-1.5}{0.45}\right) = 0.45$
$\Rightarrow \frac{t-1.5}{0.45} = -1.2816 \Rightarrow t = 0.9233$

d. A ride costing more than \$100 lasts longer than 1 hour and 50 minutes, or 1.8333 hours.
$P(X > 1.8333) = P(Z > 0.74)$
$= 1 - P(Z \leq 0.74)$
$= 1 - 0.7704 = 0.2296$

6.41 Let $X =$ the volume of timber per hectare. $X \sim N(231, 8^2)$

a. $P(225 \leq X \leq 235) = P(-0.75 \leq Z \leq 0.50)$
$= P(Z \leq 0.50) - P(Z \leq -0.75)$
$= 0.6915 - 0.2266 = 0.4649$

b. $P(230 \leq X \leq 240) = P(-0.13 \leq Z \leq 1.13)$
$= P(Z \leq 1.13) - P(Z \leq -0.13)$
$= 0.8708 - 0.4483 = 0.4225$

c. $P(X > 240 \mid X > 230) = \frac{P(X>240 \cap X>230)}{P(X>230)}$
$= \frac{P(X>240)}{P(X>230)} = \frac{1-P(Z\leq1.13)}{1-P(Z\leq-0.13)}$
$= \frac{1-0.8708}{1-0.4483} = \frac{0.1292}{0.5517} = 0.2342$

d. $P(X < 210) = P(Z < -2.63) = 0.0043$

6.43 Let $X =$ the growing time for an ear of Silver Queen corn. $X \overset{\bullet}{\sim} N(92, 5^2)$

a. $P(X < 90) = P(Z < -0.40) = 0.3446$

b. $P(95 \leq X \leq 100) = P(0.60 \leq Z \leq 1.60)$
$= P(Z \leq 1.60) - P(Z \leq 0.60)$
$= 0.9452 - 0.7257 = 0.2195$

c. $P(X \leq 95) = P(Z \leq 0.60) = 0.7257$
Let $Y =$ the number of plants ready by the 95th day. $Y \sim B(12, 0.7257)$
$P(Y = 4) = 0.0044$

d. $P(X \leq h) = 0.99 \Rightarrow P\left(Z \leq \frac{h-92}{5}\right) = 0.99$
$\Rightarrow \frac{h-92}{5} = 2.3263 \Rightarrow h = 103.6$

6.45 Let $X =$ the shoulder joint angle. $X \sim N(23.7, 1.9^2)$

a. $P(20 \leq X \leq 25) = P(-1.95 \leq Z \leq 0.68)$
$= P(Z \leq 0.68) - P(Z \leq -1.95)$
$= 0.7517 - 0.0256 = 0.7261$

b. $P(X < 18) = P(Z < -3.00) = 0.0013$

c. $P(X > 28) = P(Z > 2.26)$
$= 1 - P(Z \leq 2.26)$
$= 1 - 0.9881 = 0.0119$

d. $P(21.7 \leq X \leq 25.7) = P(-1.05 \leq Z \leq 1.05)$
$= P(Z \leq 1.05) - P(Z \leq -1.05)$

$= 0.8531 - 0.1469 = 0.7062$
Let $Y =$ the number of employees with adequate ergonomics. $Y \sim B(5, 0.7062)$
$P(Y = 4) = 0.3654$

6.47 Let $X =$ the power rating for a residential pressure washer. $X \sim N(20, \sigma^2)$

a. $P(17.5 \leq X \leq 22.5) = 0.7229$
$\Rightarrow P(X < 17.5) = 0.1386$
$\Rightarrow P\left(Z < \frac{17.5-20}{\sigma}\right) = 0.1386$
$\Rightarrow \frac{-2.5}{\sigma} = -1.0866 \Rightarrow \sigma = 2.3$

b. $P(X \geq 15) = P(Z \geq -2.17)$
$= 1 - P(Z \leq -2.17)$
$= 1 - 0.0150 = 0.9850$

c. $P(X > 26.5) = P(Z > 2.83)$
$= 1 - P(Z \leq 2.83)$
$= 1 - 0.9977 = 0.0023$

6.49 Let $X =$ the height of a dining room chair. $X \overset{\bullet}{\sim} N(85, 1.88^2)$

a. $P(X < h) = 0.99 \Rightarrow P\left(Z < \frac{h-85}{1.88}\right) = 0.99$
$\Rightarrow \frac{h-85}{1.88} = 2.3263 \Rightarrow h = 89.4$

b. $P(X > 90) = P(Z > 2.66)$
$= 1 - P(Z \leq 2.66)$
$= 1 - 0.9961 = 0.0039$

c. $P(X \leq Q_1) = 0.25 \Rightarrow P\left(Z \leq \frac{Q_1-85}{1.88}\right) = 0.25$
$\Rightarrow \frac{Q_1-85}{1.88} = -0.6745 \Rightarrow Q_1 = 83.7$

$P(X \leq Q_3) = 0.7500 \Rightarrow P\left(Z \leq \frac{Q_3-85}{1.88}\right) = 0.7500$
$\Rightarrow \frac{Q_3-85}{1.88} = 0.6745 \Rightarrow Q_3 = 86.3$

d. $P(X < 86) = 1 - 0.0718 = 0.9282$
$\Rightarrow P\left(Z < \frac{86-\mu}{1.88}\right) = 0.9282$
$\Rightarrow \frac{86-\mu}{1.88} = 1.4625 \Rightarrow \mu = 83.25$

Section 6.3: Checking the Normality Assumption

6.51 Normal probability plot:

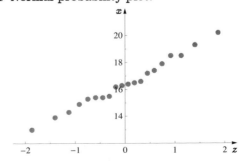

There is no evidence to suggest the data are from a

nonnormal population. The points lie along a fairly straight line.

6.53 a. There is no evidence to suggest the data are from a nonnormal population. The points appear to fall along a straight line.

b. There is evidence to suggest the data are from a nonnormal population. The points do not appear to fall on a straight line.

c. There is evidence to suggest the data are from a nonnormal population. The points do not appear to fall on a straight line.

d. There is no evidence to suggest the data are from a nonnormal population. The points appear to fall on a straight line.

6.55 Frequency histogram:

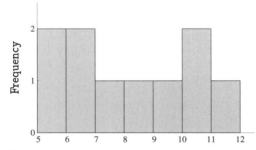

The distribution does not appear to be normal.

Backward Empirical Rule:
$\overline{x} = 8.169$; $s = 2.3427$

Interval	Proportion
Within $1s$: $(5.83, 10.51)$	0.50
Within $2s$: $(3.48, 12.85)$	1.00
Within $3s$: $(1.14, 15.20)$	1.00

The actual proportions are not close to those given by the Empirical Rule.

$IQR/s = 4.53/2.3427 = 1.9337$
The ratio is not close to 1.3.

Normal probability plot:

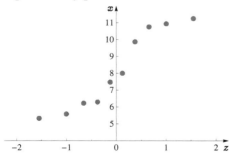

The points do not fall on a straight line. There is

evidence to suggest the data are from a nonnormal population.

6.57 Frequency histogram:

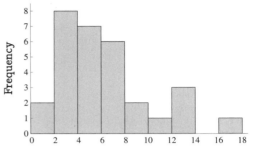

The distribution appears to be skewed to the right, and, therefore, nonnormal.

Backward Empirical Rule:
$\overline{x} = 6.2533$; $s = 3.8820$

Interval	Proportion
$(\ \ 2.37, 10.14)$	0.73
$(-1.51, 14.02)$	0.97
$(-5.39, 17.90)$	1.00

The actual proportions are close the those given by the Empirical Rule.

$IQR/s = 4.2/3.882 = 1.08$
The ratio is not close to 1.3.

Normal probability plot:

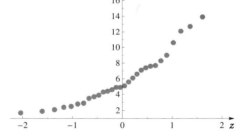

The points appear to fall along a curve.

These four methods suggest the data are from a nonnormal population.

6.59 a. $\overline{x} = \frac{1}{50}(12.70 + \cdots + 11.19) = 10.2894$
$s^2 = \frac{1}{49}(5445.2571 - \frac{1}{50}(264679.3809)) = 3.0953$
$s = \sqrt{3.0953} = 1.7593$

b. $\overline{x} \pm s = 10.2894 \pm 1.7593 = (8.53, 12.05)$
$\overline{x} \pm 2s = 10.2894 \pm 2(1.7593) = (6.77, 13.81)$
$\overline{x} \pm 3s = 10.2894 \pm 3(1.7593) = (5.01, 15.57)$

c. The proportions are 0.72, 0.96, and 1.00. There is no evidence to suggest the data are from a nonnormal population.

6.61 a. Stem-and-leaf plot:

37	3
38	36
39	
40	9
41	5
42	348
43	03446
44	025
45	6
46	36
47	
48	
49	2

Stem = 0.1
Leaf = 0.01

b. $IQR/s = 0.245/0.2865 = 0.8551$

c. Normal probability plot:

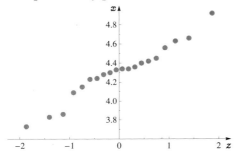

d. There is some evidence to suggest these data are from a nonnormal distribution. The stem-and-leaf plot has some outliers, the ratio IQR/s is far from 1.3, and the normal probability plot exhibits non-linearity.

6.63 Frequency histogram:

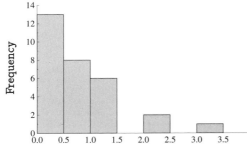

The frequency histogram is skewed to the right.

Backward Empirical Rule:
$\bar{x} = 0.7713$; $s = 0.7635$

Interval	Proportion
(0.0078, 1.5349)	0.90
(−0.7558, 2.2984)	0.93
(−1.5193, 3.0620)	0.97

These proportions do not appear to fit those given by the Empirical Rule.

$IQR/s = 0.83/0.7635 = 1.0864$

This ratio is far away from 1.3.

Normal probability plot:

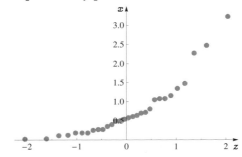

These points appear to fall along a curve.

All four techniques indicate there is evidence to suggest the data are from a nonnormal population.

Section 6.4: The Exponential Distribution

6.65 a. A graph of the density curve for X:

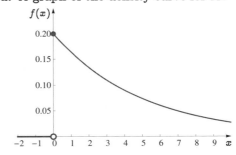

b. $\mu = 1/0.2 = 5.0$
$\sigma^2 = 1/(0.2)^2 = 25.0$, $\sigma = \sqrt{25} = 5$

c. $P(X < 3) = 1 - e^{-0.2(3)} = 0.4512$

d. $P(1 < X < 9) = P(X < 9) - P(X < 1)$
$= (1 - e^{-0.2(9)}) - (1 - e^{-0.2(1)})$
$= 0.8347 - 0.1813 = 0.6534$

6.67 a. $P(X > 30) = 1 - P(X \le 30)$
$= 1 - (1 - e^{-0.025(30)})$
$= 1 - 0.5276 = 0.4724$

Illustration:

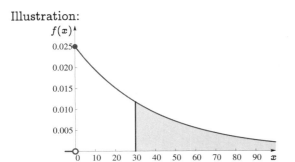

b. $P(X > 20) = 1 - P(X \leq 20)$
$= 1 - (1 - e^{-0.025(20)})$
$= 1 - 0.3935 = 0.6065$

c. $P(X > 50 \mid X \geq 30) = \frac{P(X > 50 \cap X \geq 30)}{P(X \geq 30)}$

$= \frac{P(X > 50)}{P(X \geq 30)} = \frac{1 - P(X \leq 50)}{1 - P(X < 30)}$

$= \frac{0.2865}{0.4724} = 0.6065$

6.69 $P(X_1 < 1 \cap X_2 < 1 \cap X_3 < 1 \cap X_4 < 1)$
$= P(X_1 < 1) P(X_2 < 1) P(X_3 < 1) P(X_4 < 1)$
$= (1 - e^{-0.1(1)})(1 - e^{-0.2(1)})(1 - e^{-0.3(1)})(1 - e^{-0.4(1)})$
$= (0.0952)(0.1813)(0.2592)(0.3297)$
$= 0.0015$

6.71 a. $\mu = 1/\lambda = 50,000 \Rightarrow \lambda = 0.00002$

b. $P(X > 60,000) = 1 - P(X \leq 60,000)$
$= 1 - 0.6988 = 0.3012$

c. $P(40,000 \leq X \leq 80,000)$
$= P(X \leq 80,000) - P(X \leq 40,000)$
$= 0.7981 - 0.5507 = 0.2474$

d. $P(X < 20,000) = 0.3297$

6.73 a. $P(X \leq 23) = 0.9436$

b. $P(X > 15) = 1 - P(X \leq 15)$
$= 1 - 0.8466 = 0.1534$

c. $\mu = 1/0.125 = 8$
$\sigma^2 = 1/0.125^2 = 64, \quad \sigma = \sqrt{64} = 8$

d. $P(X > t) = 0.75 \Rightarrow e^{-0.125t} = 0.75$
$\Rightarrow -0.125t = \ln 0.75 = -0.2877$
$\Rightarrow t = 2.3016$

6.75 a. $P(X > 10) = 1 - P(X \leq 10)$
$= 1 - 0.3935 = 0.6065$

b. $P(X > t) = 0.10 \Rightarrow e^{-0.05t} = 0.10$
$\Rightarrow -0.05t = \ln(0.10) = -2.3026 \Rightarrow t = 46.0517$

c. $P(X_1 > 5 \cap X_2 > 5) = P(X_1 > 5) P(X_2 > 5)$
$= (0.7788)(0.7788) = 0.6065$

6.77 a. $P(X \leq 10) = 1 - e^{-0.05(10)} = 0.3935$

b. $P(5 \leq X \leq 20) = P(X \leq 20) - P(X \leq 5)$
$= 0.6321 - 0.2212 = 0.4109$

c. $P(X > 35 \mid X > 15) = \frac{P(X > 35 \cap X > 15)}{P(X > 15)}$

$= \frac{P(X > 35)}{P(X > 15)} = \frac{1 - P(X \leq 35)}{1 - P(X \leq 15)}$

$= \frac{1 - 0.8262}{1 - 0.5276} = \frac{0.1738}{0.4724} = 0.3679$

d. $P(X_1 > 25 \cap X_2 > 25 \cap X_3 > 25 \cap X_4 > 25)$
$= P(X_1 > 25) P(X_2 > 25) P(X_3 > 25) P(X_4 > 25)$
$= (0.2865)^4 = 0.0067$

6.79 a. $\mu = 1/(1/30) = 30$
$\sigma^2 = 1/(1/30)^2 = 900, \quad \sigma = \sqrt{900} = 30$

b. $P(X \geq 40) = 1 - P(X < 40)$
$= 1 - 0.7364 = 0.2636$

c. $P(30 \leq X \leq 50) = P(X \leq 50) - P(X \leq 30)$
$= 0.8111 - 0.6321 = 0.1790$

d. $P(X \leq t) = 0.90 \Rightarrow 1 - e^{-(1/30)t} = 0.90$
$\Rightarrow e^{-(1/30)t} = 0.10 \Rightarrow -(1/30)t = \ln(0.10) =$
$-2.3026 \Rightarrow t = 69.08$

e. $P(X_1 \leq 35 \cap X_2 \leq 25) = P(X_1 \leq 35) P(X_2 \leq 25)$
$= (0.6886)(0.5654) = 0.3893$

Chapter Exercises

6.81 Let $X =$ the time until the document starts to print. X has an exponential distribution with $\lambda = 0.40$.

a. $\mu = 1/0.40 = 2.5$

b. $P(X \leq 0.5) = 1 - e^{-0.40(0.5)} = 0.1813$

c. $P(X > 5) = 1 - P(X \leq 5)$
$= 1 - 0.8647 = 0.1353$

d. $P(X > t) = 0.02 \Rightarrow e^{-0.40t} = 0.02$
$\Rightarrow -0.40t = \ln(0.02) = -3.9120 \Rightarrow t = 9.78$

6.83 Let $X =$ the length of time between a request for a review and final approval. $X \sim N(21, 16)$

a. $P(X < 14) = P(Z < -1.75) = 0.0401$

b. $P(15 \leq X \leq 19) = P(-1.5 \leq Z \leq -0.5)$
$= P(Z \leq -0.5) - P(Z \leq -1.5)$
$= 0.3085 - 0.0668 = 0.2417$

c. $P(X \leq 30 \mid X \geq 20) = \frac{P(X \leq 30 \cap X \geq 20)}{P(X \geq 20)}$

$= \frac{P(20 \leq X \leq 30)}{P(X \geq 20)} = \frac{P(-0.25 \leq Z \leq 2.25)}{P(Z \geq -0.25)}$

$= \frac{P(X \leq 2.25) - P(Z \leq -0.25)}{1 - P(Z \leq -0.25)}$

$= \frac{0.9878 - 0.4013}{1 - 0.4013} = \frac{0.5865}{0.5987} = 0.9796$

d. $P(X_1 > 30 \cap X_2 > 30) = P(X_1 > 30) P(X_2 > 30)$
$= (1 - P(X_1 \leq 30))(1 - P(X_2 \leq 30))$
$= (1 - 0.9878)(1 - 0.9878) = 0.000149$

6.85 Frequency histogram:

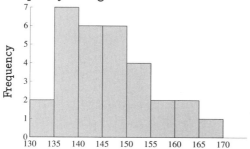

The frequency histogram is slightly skewed to the right.

Backward Empirical Rule:
$\bar{x} = 146.763$; $s = 8.6374$

Interval	Proportion
$(138.13, 155.40)$	0.70
$(129.49, 164.04)$	0.97
$(120.85, 172.68)$	1.00

These proportions are very close to those given by the Empirical Rule.

$IQR/s = 13.1/8.6374 = 1.5167$

This ratio is a little far from 1.3.

Normal probability plot:

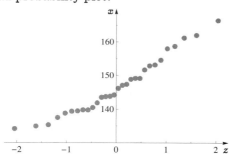

These points appear to fall along a straight line. There is a slight curve at the bottom left.

There is some evidence to suggest the data are from a nonnormal population. The histogram is positively skewed, IQR/s is not close to 1.3, and the normal probability plot has a slight arc.

6.87 Let X = the thickness of the celluloid base. $X \sim N(0.125, 0.025^2)$

a. $P(0.10 \leq X \leq 0.17) = P(-1.00 \leq Z \leq 1.80)$
$= P(Z \leq 1.80) - P(Z \leq -1.00)$
$= 0.9641 - 0.1587 = 0.8054$

b. $P(X > 0.20) = P(Z > 3.00)$
$= 1 - P(Z \leq 3.00)$
$= 1 - 0.9987 = 0.0013$

c. $P(X < 0.04) = P(Z < -3.40) = 0.0003$

d. $P(X_1 > 0.15 \cap X_2 > 0.15 \cap X_3 > 0.15)$
$= P(X_1 > 0.15)\, P(X_2 > 0.15)\, P(X_3 > 0.15)$
$= P(Z > 1)^3 = (1 - 0.8413)^3 = 0.0040$

6.89 Frequency histogram:

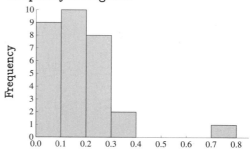

The frequency histogram is slightly skewed to the right.

Backward Empirical Rule:
$\bar{x} = 0.173$, $s = 0.1289$

Interval	Proportion
$(0.0441, 0.3019)$	0.87
$(-0.0847, 0.4307)$	0.97
$(-0.2136, 0.5596)$	0.97

These proportions are not very close to those given by the Empirical Rule.

$IQR/s = 0.16/0.1289 = 1.2413$

This ratio is close to 1.3.

Normal probability plot:

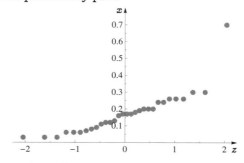

Most of the points fall along a straight line. There appears to be one outlier.

There is some evidence the data are from a nonnormal population. The histogram and the normal probability plot indicate an outlier, and the backward Empirical Rule proportions are inconsistent.

6.91 Let X = the amount of caffeine in an 8-ounce cup of Maxwell House coffee. $X \overset{\bullet}{\sim} N(110, 7.75^2)$

a. $P(X < 100) = P(Z < -1.29) = 0.0985$

b. $P(105 \leq X \leq 120) = P(-0.65 \leq Z \leq 1.29)$
$= P(Z \leq 1.29) - P(Z \leq -0.65)$
$= 0.9015 - 0.2578 = 0.6437$

c. $P(X > 130) = P(Z > 2.58)$
$= 1 - P(Z \leq 2.58)$
$= 1 - 0.9951 = 0.0049$

6.93 Let $X =$ the amount of a new car loan.
$X \overset{\bullet}{\sim} N(17{,}000, 4500^2)$

a. $P(X \leq 10{,}000) = P(Z \leq -1.56) = 0.0594$

b. $P(15{,}000 \leq X \leq 20{,}000) = P(-0.44 \leq Z \leq 0.67)$
$= P(Z \leq 0.67) - P(Z \leq -0.44)$
$= 0.7486 - 0.3300 = 0.4186$

c. $P(17{,}000 - a \leq X \leq 17{,}000 + a) = 0.95$
$\Rightarrow P(X < 17{,}000 - a) = 0.025$
$\Rightarrow P(Z < -a/4500) = 0.025$
$\Rightarrow -a/4500 = -1.96 \Rightarrow a = 8820$
Interval: $(8180, 25820)$

d. Let $Y =$ the length of a new car loan.
$Y \overset{\bullet}{\sim} N(62, 16)$

$P(X > 25{,}000 \cap Y < 56)$
$= P(X > 25{,}000) \, P(Y < 56)$
$= P(Z > 1.78) \, P(Z < -1.50)$
$= (1 - P(Z \leq 1.78)) \cdot P(Z < -1.50)$
$= (1 - 0.9625)(0.0668) = 0.0025$

6.95 Let $X =$ the proportion of water in a randomly selected barrel of oil pumped to the surface. $X \sim N(0.12, 0.025^2)$

a. $P(X < 0.12) = P(Z < 0) = 0.50$

b. $P(0.15 \leq X \leq 0.17) = P(1.20 \leq Z \leq 2.00)$
$= P(Z \leq 2.00) - P(Z \leq 1.20)$
$= 0.9772 - 0.8849 = 0.0923$

c. $P(X > 0.20) = P(Z > 3.20)$
$= 1 - P(Z \leq 3.20)$
$= 1 - 0.9993 = 0.0007$

6.97 a. Probability density function:

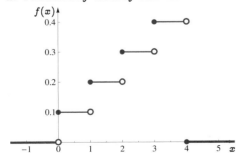

b. $P(X < 1.5) = (1)(0.1) + (0.5)(0.2) = 0.2$

c. $P(X > 3) = (1)(0.4) = 0.4$

d. $P(2 \leq X \leq 4) = (1)(0.3) + (1)(0.4) = 0.7$

e. $P(X \leq 1 \mid X \leq 3) = \frac{P(X \leq 1 \cap X \leq 3)}{P(X \leq 3)}$

$= \frac{P(X \leq 1)}{P(X \leq 3)} = \frac{(1)(0.1)}{(1)(0.1) + (1)(0.2) + (1)(0.3)}$

$= \frac{0.1}{0.6} = 0.1667$

6.99 a. Probability density function:

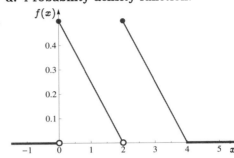

b. $P(X < 1) = \frac{1}{2}(1)(0.5 + 0.25) = 0.375$

c. $P(X \geq 3) = \frac{1}{2}(1)(0.25) = 0.125$

d. $P(1 < X < 3)$
$= \frac{1}{2}(1)(0.25) + \frac{1}{2}(1)(0.5 + 0.25) = 0.50$

Chapter 7: Sampling Distributions

Section 7.1: Statistics, Parameters, and Sampling Distributions

7.1 a. Statistic   **b.** Parameter   **c.** Statistic   **d.** Parameter   **e.** Statistic

7.3 a. $\mu = \frac{1}{5}(10 + \cdots + 25) = 16$, $\tilde{\mu} = 15$

b. Table of all possible samples, the computed value of the sample mean, and the probability of each sample:

Sample	$\bar{x}$	Prob.	Sample	$\bar{x}$	Prob.
10, 12, 15	12.33	0.10	10, 12, 18	13.33	0.10
10, 12, 25	15.67	0.10	10, 15, 18	14.33	0.10
10, 15, 25	16.67	0.10	10, 18, 25	17.67	0.10
12, 15, 18	15.00	0.10	12, 15, 25	17.33	0.10
12, 18, 25	18.33	0.10	15, 18, 25	19.33	0.10

Sampling distribution:

$\bar{x}$	12.33	13.33	14.33	15.00	15.67
$p(\bar{x})$	0.1	0.1	0.1	0.1	0.1

$\bar{x}$	16.67	17.33	17.67	18.33	19.33
$p(\bar{x})$	0.1	0.1	0.1	0.1	0.1

$\mu_{\bar{X}} = (12.33)(0.1) + \cdots + (19.33)(0.1) = 16$

$\sigma_{\bar{X}}^2 = 260.57 - 16^2 = 4.6$

$\sigma_{\bar{X}} = \sqrt{4.6} = 2.1448$

c. Table of all possible samples, the computed value of the sample median, and the probability of each sample:

Sample	$\tilde{x}$	Prob.	Sample	$\tilde{x}$	Prob.
10, 12, 15	12	0.10	10, 12, 18	12	0.10
10, 12, 25	12	0.10	10, 15, 18	15	0.10
10, 15, 25	15	0.10	10, 18, 25	18	0.10
12, 15, 18	15	0.10	12, 15, 25	15	0.10
12, 18, 25	18	0.10	15, 18, 25	18	0.10

Sampling distribution:

$\tilde{x}$	12	15	18
$p(\tilde{x})$	0.3	0.4	0.3

$\mu_{\tilde{X}} = (12)(0.3) + (15)(0.4) + (18)(0.3) = 15$

$\sigma_{\tilde{X}}^2 = 230.4 - 15^2 = 5.4$

$\sigma_{\tilde{X}} = \sqrt{5.4} = 2.3238$

d. The mean of the sample mean is the population mean. The mean of the sample median is the population median.

7.5 a. Random samples will vary.

b. Frequency histogram:

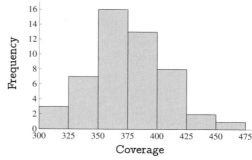

c. The sampling distribution is approximately normal. The mean of the sample distribution is approximately 375.

d. The population mean is $\mu = 379.7$, which is almost the same as the mean of the sampling distribution.

7.7 a. Table of all possible samples, the computed value of the sample median, and the probability of each sample:

Sample	$\tilde{x}$	Prob.	Sample	$\tilde{x}$	Prob.
0, 0	0.0	0.2500	0, 1	0.5	0.1250
0, 2	1.0	0.0500	0, 3	1.5	0.0500
0, 4	2.0	0.0250	1, 0	0.5	0.1250
1, 1	1.0	0.0625	1, 2	1.5	0.0250
1, 3	2.0	0.0250	1, 4	2.5	0.0125
2, 0	1.0	0.0500	2, 1	1.5	0.0250
2, 2	2.0	0.0100	2, 3	2.5	0.0100
2, 4	3.0	0.0050	3, 0	1.5	0.0500
3, 1	2.0	0.0250	3, 2	2.5	0.0100
3, 3	3.0	0.0100	3, 4	3.5	0.0050
4, 0	2.0	0.0250	4, 1	2.5	0.0125
4, 2	3.0	0.0050	4, 3	3.5	0.0050
4, 4	4.0	0.0025			

Distribution of the sample median:

$\tilde{x}$	0.0	0.5	1.0	1.5	2.0
$p(\tilde{x})$	0.2500	0.2500	0.1625	0.1500	0.1100

$\tilde{x}$	2.5	3.0	3.5	4.0
$p(\tilde{x})$	0.0450	0.0200	0.0100	0.0025

b. $\mu_{\tilde{X}} = (0)(0.25) + \cdots + (4)(0.0025) = 0.95$

$\sigma_{\tilde{X}}^2 = 1.6263 - (0.95)^2 = 0.7238$

$\sigma_{\tilde{X}} = \sqrt{0.7238} = 0.8508$

7.9 a. Table of all possible samples, the computed value of the sample mean, and the probability of each sample:

Sample	$\bar{x}$	Prob.	Sample	$\bar{x}$	Prob.
9.9, 9.7	9.80	0.1	9.9, 7.1	8.50	0.1
9.9, 7.0	8.45	0.1	9.9, 6.8	8.35	0.1
9.7, 7.1	8.40	0.1	9.7, 7.0	8.35	0.1
9.7, 6.8	8.25	0.1	7.1, 7.0	7.05	0.1
7.1, 6.8	6.95	0.1	7.0, 6.8	6.90	0.1

Distribution of the sample mean:

$\bar{x}$	6.90	6.95	7.05	8.25	8.35
$p(\bar{x})$	0.1	0.1	0.1	0.1	0.2

$\bar{x}$	8.40	8.45	8.50	9.80
$p(\bar{x})$	0.1	0.1	0.1	0.1

b. Table of all possible samples, the computed value of the sample total, and the probability of each sample:

Sample	t	Prob.	Sample	t	Prob.
9.9, 9.7	19.6	0.1	9.9, 7.1	17.0	0.1
9.9, 7.0	16.9	0.1	9.9, 6.8	16.7	0.1
9.7, 7.1	16.8	0.1	9.7, 7.0	16.7	0.1
9.7, 6.8	16.5	0.1	7.1, 7.0	14.1	0.1
7.1, 6.8	13.9	0.1	7.0, 6.8	13.8	0.1

Distribution of the sample total:

t	13.8	13.9	14.1	16.5	16.7
$p(t)$	0.1	0.1	0.1	0.1	0.2

t	16.8	16.9	17.0	19.6
$p(t)$	0.1	0.1	0.1	0.1

7.11 a. $\mu = (5)(0.50) + \cdots + (8)(0.05) = 5.75$

$\sigma^2 = 33.85 - 5.75^2 = 0.7875$

b. Table of all possible samples, the computed value of the sample variance, and the probability of each sample:

Sample	s^2	Prob.	Sample	s^2	Prob.
5, 5	0.0	0.2500	5, 6	0.5	0.1500
5, 7	2.0	0.0750	5, 8	4.5	0.0250
6, 5	0.5	0.1500	6, 6	0.0	0.0900
6, 7	0.5	0.0450	6, 8	2.0	0.0150
7, 5	2.0	0.0750	7, 6	0.5	0.0450
7, 7	0.0	0.0225	7, 8	0.5	0.0075
8, 5	4.5	0.0250	8, 6	2.0	0.0150
8, 7	0.5	0.0075	8, 8	0.0	0.0025

Distribution of the sample variance:

s^2	0.0	0.5	2.0	4.5
$p(s^2)$	0.365	0.405	0.180	0.050

c. $\mu_{S^2} = (0)(0.365) + \cdots + (4.5)(0.050) = 0.7875$
This is the same as the variance of X.

7.13 a. Table of all possible samples, the computed value of the sample maximum, and the probability of each sample:

Sample	m	Prob.	Sample	m	Prob.
83, 100	100	0.1667	83, 95	95	0.1667
83, 70	83	0.1667	100, 95	100	0.1667
100, 70	100	0.1667	95, 70	95	0.1667

Distribution of the maximum weight:

m	83	95	100
$p(m)$	0.1667	0.3333	0.5000

b. Table of all possible samples, the computed value of the sample total, and the probability of each sample:

Sample	t	Prob.	Sample	t	Prob.
83, 100	183	0.1667	83, 95	178	0.1667
83, 70	153	0.1667	100, 95	195	0.1667
100, 70	170	0.1667	95, 70	165	0.1667

Distribution of the total weight:

t	153	165	170	178	183	195
$p(t)$	0.1667	0.1667	0.1667	0.1667	0.1667	0.1667

7.15 a. $\mu = \frac{1}{50}(1 + 2 + \cdots + 50) = 25.5$

b. Random samples will vary.

c. Frequency histogram:

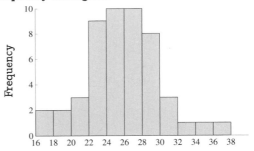

The histogram is centered near the population mean, 25.5.

Section 7.2: The Sampling Distribution of the Sample Mean

7.17 a. $\overline{X} \sim N(10, 6.25/7)$
$P(\overline{X} \le 9) = P(Z \le -1.06) = 0.1446$

b. $\overline{X} \sim N(10, 6.25/12)$
$P(\overline{X} > 11.5) = P(Z > 1.59)$
$= 1 - P(Z \le 2.08)$
$= 1 - 0.9812 = 0.0188$

c. $\overline{X} \sim N(10, 6.25/15)$
$P(9.5 \leq \overline{X} \leq 10.5) = P(-0.77 \leq Z \leq 0.77)$
$= P(Z \leq 0.77) - P(Z \leq -0.77)$
$= 0.7794 - 0.2206 = 0.5588$

d. $\overline{X} \sim N(10, 6.25/25)$
$P(\overline{X} \geq 10.25) = P(Z \geq 0.50)$
$= 1 - P(Z \leq 0.50)$
$= 1 - 0.6915 = 0.3085$

e. $\overline{X} \sim N(10, 6.25/100)$
$P(\overline{X} \leq 9.8 \cup \overline{X} \geq 10.2)$
$= 2P(\overline{X} \leq 9.8) = 2P(Z \leq -0.8)$
$= 2(0.2119) = 0.4238$

7.19 a. $\overline{X} \overset{\bullet}{\sim} N(50, 49/38)$. The Central Limit Theorem is necessary because the shape of the underlying distribution is not known.

b. $P(\overline{X} < 49) = P(Z < -0.88) = 0.1894$

c. $P(\overline{X} \geq 52) = P(Z \geq 1.76)$
$= 1 - P(Z \leq 1.76)$
$= 1 - 0.9608 = 0.0392$

d. $P(49.5 \leq \overline{X} \leq 51.5) = P(-0.44 \leq Z \leq 1.32)$
$= P(Z \leq 1.32) - P(Z \leq -0.44)$
$= 0.9066 - 0.3300 = 0.5766$

e. $P(\overline{X} > c) = 0.15 \Rightarrow P(\overline{X} \leq c) = 0.85$
$\Rightarrow P\left(Z \leq \frac{c-50}{7/\sqrt{38}}\right) = 0.85$

$\Rightarrow \frac{c-50}{7/\sqrt{38}} = 1.0364 \Rightarrow c = 51.1769$

7.21 a. $\overline{X} \overset{\bullet}{\sim} N(30, 50^2/40)$
Graph of the density curve $y = f(\overline{x})$:

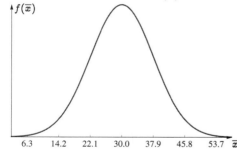

b. $P(\overline{X} \geq 38) = P(Z \geq 1.01)$
$= 1 - P(Z \leq 1.01)$
$= 1 - 0.8438 = 0.1562$

c. $P(20 \leq \overline{X} \leq 40) = P(-1.26 \leq Z \leq 1.26)$
$= P(Z \leq 1.26) - P(Z \leq -1.26)$
$= 0.8962 - 0.1038 = 0.7924$

d. $P(\overline{X} < 15) = P(Z < -1.90) = 0.0287$

e. $P(\overline{X} \leq c) = 0.001 \Rightarrow P\left(Z \leq \frac{c-30}{50/\sqrt{40}}\right) = 0.001$

$\Rightarrow \frac{c-30}{50/\sqrt{40}} = -3.0902 \Rightarrow c = 5.57$

7.23 Blue: graph of the probability density function for X. Green: graph of the probability density function for $\overline{X}$ with $n = 5$. Red: graph of the probability density function for $\overline{X}$ with $n = 15$.

7.25 a. $\overline{X} \overset{\bullet}{\sim} N(8.25, 0.01/35)$
Graph of the density curve $y = f(\overline{x})$:

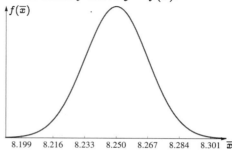

b. $P(\overline{X} > 8) = P(Z > -14.79) \approx 1$

c. $P(8.2 \leq \overline{X} \leq 8.4) = P(-2.96 \leq Z \leq 8.87)$
$= P(Z \leq 8.87) - P(Z \leq -2.96)$
$= 1 - 0.0015 = 0.9985$

d. For $\sigma = 0.15$, $\overline{X} \overset{\bullet}{\sim} N(8.25, 0.15^2/35)$
Graph of the density curve $y = f(\overline{x})$:

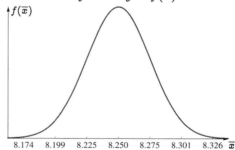

$P(\overline{X} > 8) = P(Z > -9.86) \approx 1$
$P(8.2 \leq \overline{X} \leq 8.4) = P(-1.97 \leq Z \leq 5.92)$
$= P(Z \leq 5.92) - P(Z \leq -1.97)$
$= 1 - 0.0244 = 0.9756.$

7.27 $\overline{X} \overset{\bullet}{\sim} N(100, 12^2/40)$

a. $P(\overline{X} > 102) = P(Z > 1.05)$
$= 1 - P(Z \leq 1.05)$
$= 1 - 0.8531 = 0.1469$

b. $P(101 \leq \overline{X} \leq 103) = P(0.53 \leq Z \leq 1.58)$
$= P(Z \leq 1.58) - P(Z \leq 0.53)$
$= 0.9429 - 0.7019 = 0.2410$

c. Claim: $\mu = 100 \Rightarrow \overline{X} \overset{\bullet}{\sim} N(100, 144/40)$
Experiment: $\overline{x} = 98.5$
Likelihood: $P(\overline{X} \leq 98.5) = P(Z \leq -0.79) = 0.2148$
Conclusion: There is no evidence to suggest the claim is false, that μ is less than 100.

7.29 a. $\overline{X} \stackrel{\bullet}{\sim} N(6.5, 4^2/35)$

b. $P(\overline{X} > 7) = P(Z > 0.74)$
$= 1 - P(Z \leq 0.74)$
$= 1 - 0.7704 = 0.2296$

c. $\mu = 6.5 \Rightarrow \overline{X} \stackrel{\bullet}{\sim} N(6.5, 16/35)$
Experiment: $\overline{x} = 5.1$
Likelihood: $P(\overline{X} \leq 5.1) = P(Z \leq -2.07) = 0.0192$
Conclusion: There is evidence to suggest the claim is false, that the mean police standoff time is lower.

7.31 $\overline{X} \stackrel{\bullet}{\sim} N(115.054, 30^2/50)$

a. $P(\overline{X} < 110) = P(Z < -1.19) = 0.1170$

b. $P(115 \leq \overline{X} \leq 120) = P(-0.01 \leq Z \leq 1.17)$
$= P(Z \leq 1.17) - P(Z \leq -0.01)$
$= 0.8790 - 0.4960 = 0.3830$

c. $P(\overline{X} > 125) = P(Z > 2.34)$
$= 1 - P(Z \leq 2.34)$
$= 1 - 0.9904 = 0.0096$

7.33 a. $\overline{X} \sim N(4.125, 1/15)$

b. $P(\overline{X} > 4.5) = P(Z > 1.45)$
$= 1 - P(Z \leq 1.45)$
$= 1 - 0.9265 = 0.0735$

c. $P(3.375 \leq \overline{X} \leq 4.125) = P(-2.90 \leq Z \leq 0)$
$= P(Z \leq 0) - P(Z \leq -2.9)$
$= 0.5000 - 0.0019 = 0.4981$

d. $P(\overline{X} < 3.5) = P(Z < -2.42) = 0.0078$

7.35 $\overline{X} \stackrel{\bullet}{\sim} N(320, 35^2/49)$

a. $P(\overline{X} > 330) = P(Z > 2.00)$
$= 1 - P(Z \leq 2.00)$
$= 1 - 0.9772 = 0.0228$

b. $P(315 \leq \overline{X} \leq 325) = P(-1.00 \leq Z \leq 1.00)$
$= P(Z \leq 1.00) - P(Z \leq -1.00)$
$= 0.8413 - 0.1587 = 0.6826$

c. Claim: $\mu = 320 \Rightarrow \overline{X} \stackrel{\bullet}{\sim} N(320, 25)$
Experiment: $\overline{x} = 310$
Likelihood: $P(\overline{X} \leq 310) = P(Z \leq -2.00) = 0.0228$
Conclusion: There is evidence to suggest the claim is false, that the mean ozone-layer thickness is less than 320 DU.

7.37 a. $T \stackrel{\bullet}{\sim} N(35 \cdot 15, 35 \cdot 4) = N(525, 140)$

b. $P(480 \leq T \leq 540) = P(-3.80 \leq Z \leq 1.27)$
$= P(Z \leq 1.27) - P(Z \leq -3.80)$
$= 0.8980 - 0.0001 = 0.8979$

c. $P(T > 552) = P(Z > 2.28)$
$= 1 - P(Z \leq 2.28)$
$= 1 - 0.9887 = 0.0113$

d. $P(T \geq t) = 0.01 \Rightarrow P(T < t) = 0.99$
$\Rightarrow P\left(Z < \frac{t-525}{\sqrt{140}}\right) = 0.99$
$\Rightarrow \frac{t-525}{\sqrt{140}} = 2.3263 \Rightarrow t = 552.53$

7.39 Let $X =$ the force required. $\overline{X} \sim N(\mu, 0.1^2/25)$

a. $P(\overline{X} < 0.75) + P(\overline{X} > 0.85)$
$= 2P(Z < -2.50) = 2(0.0062) = 0.0124$

b. For $\mu = 0.82$:
$P(0.75 \leq \overline{X} \leq 0.85) = P(-3.50 \leq Z \leq 1.50)$
$= P(Z \leq 1.50) - P(Z \leq -3.50)$
$= 0.9332 - 0.0002 = 0.9330$

For $\mu = 0.84$:
$P(0.75 \leq \overline{X} \leq 0.85) = P(-4.50 \leq Z \leq 0.50)$
$= P(Z \leq 0.50) - P(Z \leq -4.50)$
$= 0.6915 - 0.0000 = 0.6915$

c. $P(\overline{X} < 0.76) + P(\overline{X} > 0.84)$
$= 2P(Z < -2.00) = 2(0.0228) = 0.0456$

For $\mu = 0.82$:
$P(0.76 \leq \overline{X} \leq 0.84) = P(-3.00 \leq Z \leq 1.00)$
$= P(Z \leq 1.00) - P(Z \leq -3.00)$
$= 0.8413 - 0.0013 = 0.8400$

For $\mu = 0.84$:
$P(0.76 \leq \overline{X} \leq 0.84) = P(-4.00 \leq Z \leq 0.00) \approx 0.50$

7.41 a. Table of the population mean and the probability of accepting the entire shipment:

μ	Prob	μ	Prob
1.86	0.0001	1.88	0.0016
1.90	0.0174	1.92	0.1038
1.94	0.3372	1.96	0.6627
1.98	0.8946	2.00	0.9652
2.02	0.8946	2.04	0.6627
2.06	0.3372	2.08	0.1038
2.10	0.0174	2.12	0.0016
2.14	0.0001		

b. Graph of the OC curve:

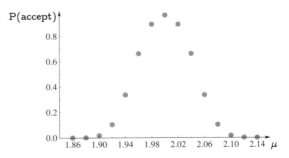

7.43 a. $\mu = \frac{1}{20}(30 + \cdots + 30) = 30.85$

b. Answers will vary.

c. Histogram of the sample means:

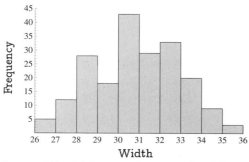

The shape of the histogram is approximately normal, centered near the population mean 30.85.

Section 7.3: The Distribution of the Sample Proportion

7.45 $\widehat{P} \overset{\bullet}{\sim} N(0.40, 0.0012)$

a. $P(\widehat{P} \leq 0.37) = P(Z \leq -0.87) = 0.1922$

b. $P(\widehat{P} > 0.45) = P(Z > 1.44)$
$= 1 - P(Z \leq 1.44)$
$= 1 - 0.9251 = 0.0749$

c. $P(0.38 \leq \widehat{P} \leq 0.42) = P(-0.58 \leq Z \leq 0.58)$
$= P(Z \leq 0.58) - P(Z \leq -0.58)$
$= 0.7190 - 0.2810 = 0.4380$

d. $P(\widehat{P} < 0.33)$ or $P(\widehat{P} > 0.47)$
$= 2P(\widehat{P} < 0.33) = 2P(Z < -2.02)$
$= 2(0.0217) = 0.0434$

7.47 $\widehat{P} \overset{\bullet}{\sim} N(0.35, 0.0028)$

a. $P(\widehat{P} \leq a) = 0.10 \Rightarrow P\left(Z \leq \frac{a-0.35}{\sqrt{0.0028}}\right) = 0.10$
$\Rightarrow \frac{a-0.35}{\sqrt{0.0028}} = -1.2816 \Rightarrow a = 0.2817$

b. $P(\widehat{P} > b) = 0.01 \Rightarrow P\left(Z \leq \frac{b-0.35}{\sqrt{0.0028}}\right) = 0.99$
$\Rightarrow \frac{b-0.35}{\sqrt{0.0028}} = 2.3263 \Rightarrow b = 0.4741$

c. $P(0.35 - c \leq \widehat{P} \leq 0.35 + c) = 0.95$
$\Rightarrow P(\widehat{P} \leq 0.35 - c) = 0.025$
$\Rightarrow P(Z \leq -c/\sqrt{0.0028}) = 0.025$
$\Rightarrow -c/\sqrt{0.0028} = -1.96 \Rightarrow c = 0.1045$

7.49 a. $\widehat{P} \overset{\bullet}{\sim} N(0.70, 0.00084)$

b. $P(\widehat{P} < 0.66) = P(Z < -1.38) = 0.0838$

c. $P(\widehat{P} > 0.71) = P(Z > 0.35)$
$= 1 - P(Z \leq 0.35)$
$= 1 - 0.6368 = 0.3632$

d. $P(0.68 \leq \widehat{P} \leq 0.78) = P(-0.69 \leq Z \leq 2.76)$
$= P(Z \leq 2.76) - P(Z \leq -0.68)$
$= 0.9971 - 0.2451 = 0.7520$

7.51 a. $\widehat{P} \overset{\bullet}{\sim} N(0.48, 0.00208)$

b. $P(\widehat{P} > 0.50) = P(Z > 0.44)$
$= 1 - P(Z \leq 0.44)$
$= 1 - 0.6700 = 0.3300$

c. $P(\widehat{P} < 0.40) = P(Z < -1.75) = 0.0401$

d. $P(0.48 - c \leq \widehat{P} \leq 0.48 + c) = 0.90$
$\Rightarrow P(\widehat{P} \leq 0.48 - c) = 0.05$
$\Rightarrow P(Z \leq -c/\sqrt{0.00208}) = 0.05$
$\Rightarrow -c/\sqrt{0.00208} = -1.6449 \Rightarrow c = 0.075$
Interval: $(0.4050, 0.5550)$

7.53 $\widehat{P} \overset{\bullet}{\sim} N(0.36, 0.001152)$

a. $P(\widehat{P} < 0.35) = P(Z < -0.29) = 0.3859$

b. $P(\widehat{P} > 0.40) = P(Z > 1.18)$
$= 1 - P(Z \leq 1.18)$
$= 1 - 0.8810 = 0.1190$

c. $P(\widehat{P} < t) = 0.95$
$\Rightarrow P\left(Z < \frac{t-0.36}{\sqrt{0.001152}}\right) = 0.95$
$\Rightarrow \frac{t-0.36}{\sqrt{0.001152}} = 1.6449 \Rightarrow t = 0.4158$

7.55 $\widehat{P} \overset{\bullet}{\sim} N(0.40, 0.0024)$

a. $P(\widehat{P} < 0.42) = P(Z < 0.41) = 0.6591$

b. $P(0.30 \leq \widehat{P} \leq 0.35) = P(-2.04 \leq Z \leq -1.02)$
$= P(Z \leq -1.02) - P(Z \leq -2.04)$
$= 0.1539 - 0.0207 = 0.1332$

c. Claim: $p = 0.40 \Rightarrow \widehat{P} \overset{\bullet}{\sim} N(0.40, 0.0024)$
Experiment: $\widehat{p} = 0.47$
Likelihood: $P(\widehat{P} \geq 0.47) = 0.0764$
Conclusion: There is no evidence to suggest the claim is false, that the acceptance rate has increased.

7.57 a. $\widehat{P} \overset{\bullet}{\sim} N(0.005, 0.000004975)$

b. $P(\widehat{P} < 0.002) = P(Z < -1.35) = 0.0885$

c. $P(\widehat{P} > 0.010) = P(Z > 2.24)$
$= 1 - P(Z \leq 2.24)$
$= 1 - 0.9875 = 0.0125$

d. $P(\widehat{P} \geq a) = 0.80 \Rightarrow P(\widehat{P} \leq a) = 0.20$
$\Rightarrow P\left(Z \leq \frac{a-0.005}{\sqrt{0.00000495}}\right) = 0.20$
$\Rightarrow \frac{a-0.005}{\sqrt{0.000004975}} = -0.8416 \Rightarrow a = 0.0031$

7.59 $\widehat{P} \overset{\bullet}{\sim} N(0.272, 0.0004605)$

a. $P(\widehat{P} > 0.31) = P(Z > 1.77)$
$= 1 - P(Z \leq 1.77)$
$= 1 - 0.9616 = 0.0384$

b. $P(0.25 \leq \widehat{P} \leq 0.30) = P(-1.03 \leq Z \leq 1.30)$
$= P(Z \leq 1.30) - P(Z \leq -1.03)$
$= 0.9032 - 0.1515 = 0.7517$

c. Claim: $p = 0.272 \Rightarrow \widehat{P} \overset{\bullet}{\sim} N(0.272, 0.0004605)$
Experiment: $\widehat{p} = 115/430 = 0.2674$
Likelihood:
$P(\widehat{P} \leq 0.2674) = P(Z \leq -0.21) = 0.4168$
Conclusion: There is no evidence to suggest the claim is false, that the true proportion of cigarette debris has changed.

7.61 a. $p = 0.05 \Rightarrow \widehat{P} \overset{\bullet}{\sim} N(0.05, 0.0002375)$
$P(\widehat{P} > 0.09) = P(Z > 2.60)$
$= 1 - P(Z \leq 2.60)$
$= 1 - 0.9953 = 0.0047$

b. $p = 0.03 \Rightarrow \widehat{P} \overset{\bullet}{\sim} N(0.03, 0.0001455)$
$P(\widehat{P} > 0.09) = P(Z > 4.97) \approx 0$

c. $p = 0.10 \Rightarrow \widehat{P} \overset{\bullet}{\sim} N(0.10, 0.00045)$
$P(\widehat{P} \leq 0.09) = P(Z \leq -0.47) = 0.3192$

7.63 For $n = 100$, here is a plot of $\sigma_{\widehat{P}}^2$ versus p:

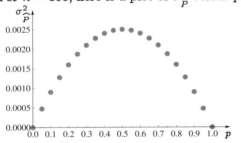

This graph suggests that the variance is a maximum when $p = 0.50$.

Chapter Exercises

7.65 Claim: $\mu = 0.5 \Rightarrow \overline{X} \overset{\bullet}{\sim} N(0.5, 0.0008)$
Experiment: $\overline{x} = 0.6$
Likelihood: $P(\overline{X} \geq 0.6) = 0.0002$
Conclusion: There is evidence to suggest the claim is false, that the mean coefficient of static friction is greater than 0.5.

7.67 a. Table of all possible samples, the computed value of the sample maximum, and the probability of each sample:

Sample	m	Prob.	Sample	m	Prob.
1, 1	1	0.0004	1, 2	2	0.0060
1, 3	3	0.0020	1, 4	4	0.0060
1, 5	5	0.0040	1, 6	6	0.0016
2, 1	2	0.0060	2, 2	2	0.0900
2, 3	3	0.0300	2, 4	4	0.0900
2, 5	5	0.0600	2, 6	6	0.0240
3, 1	3	0.0020	3, 2	3	0.0300
3, 3	3	0.0100	3, 4	4	0.0300
3, 5	5	0.0200	3, 6	6	0.0080
4, 1	4	0.0060	4, 2	4	0.0900
4, 3	4	0.0300	4, 4	4	0.0900
4, 5	5	0.0600	4, 6	6	0.0240
5, 1	5	0.0040	5, 2	5	0.0600
5, 3	5	0.0200	5, 4	5	0.0600
5, 5	5	0.0400	5, 6	6	0.0160
6, 1	6	0.0016	6, 2	6	0.0240
6, 3	6	0.0080	6, 4	6	0.0240
6, 5	6	0.0160	6, 6	6	0.0064

Distribution of M:

m	1	2	3	4	5	6
$p(m)$	0.0004	0.1020	0.0740	0.3420	0.3280	0.1536

b. $\mu_M = (1)(0.0004) + \cdots + (6)(0.1536) = 4.356$
$\sigma_M^2 = 20.276 - (4.356)^2 = 1.3013$
$\sigma_M = \sqrt{1.3013} = 1.1407$

7.69 $\overline{X} \overset{\bullet}{\sim} N(7.5, 1.75^2/35)$

a. $P(\overline{X} < 7) = P(Z < -1.69) = 0.0455$

b. $P(7.25 \leq \overline{X} \leq 7.5) = P(-0.85 \leq Z \leq 0.00)$
$= P(Z \leq 0.00) - P(Z \leq -0.85)$
$= 0.5000 - 0.1977 = 0.3023$

c. Claim: $\mu = 7.5 \Rightarrow \overline{X} \overset{\bullet}{\sim} N(7.5, 0.0875)$
Experiment: $\overline{x} = 8.1$
Likelihood: $P(\overline{X} \geq 8.1) = 0.0212$
Conclusion: There is evidence to suggest the claim is false, that the mean oxygen produced is greater than 7.5.

d. $P(\overline{X} < 7) = P(Z < -0.79) = 0.2148$
$P(7.25 \leq \overline{X} \leq 7.5) = P(-0.39 \leq Z \leq 0.00)$
$= P(Z \leq 0.00) - P(Z \leq -0.39)$
$= 0.5000 - 0.3483 = 0.1517$
Claim: $\mu = 7.5 \Rightarrow \overline{X} \overset{\bullet}{\sim} N(7.5, 0.4018)$
Experiment: $\overline{x} = 8.1$
Likelihood: $P(\overline{X} \geq 8.1) = 0.1711$
Conclusion: There is no evidence to suggest the claim is false.

7.71 $\overline{X} \overset{\bullet}{\sim} N(155, 35^2/30)$

a. $P(\overline{X} > 170) = P(Z > 2.35)$
$= 1 - P(Z \leq 2.35)$
$= 1 - 0.9906 = 0.0094$

b. $P(140 \leq \overline{X} \leq 150) = P(-2.35 \leq Z \leq -0.78)$
$= P(Z \leq -0.78) - P(Z \leq -2.35)$
$= 0.2177 - 0.0094 = 0.2083$

c. $P(\overline{X} < r) = 0.001 \Rightarrow P\left(Z < \frac{r-155}{35/\sqrt{30}}\right) = 0.001$

$\Rightarrow \frac{r-155}{35/\sqrt{30}} = -3.0902 \Rightarrow r = 135.25$

7.73 a. Statistic   **b.** Parameter   **c.** Statistic
d. Parameter   **e.** Statistic

7.75 a. f_1: underlying distribution; f_2:
distribution of the sample mean.

b. f_1: underlying distribution; f_2: distribution of
the sample mean.

c. f_1: distribution of the sample mean; f_1:
underlying distribution.

d. f_1: underlying distribution; f_2: distribution of
the sample mean.

7.77 a. $\widehat{P} \overset{\bullet}{\sim} N(0.12, 0.0004224)$

Graph of the density curve $y = f(\widehat{p})$:

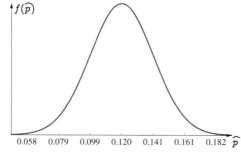

b. $P(\widehat{P} > 0.15) = P(Z > 1.46)$
$= 1 - P(Z \leq 1.46)$
$= 1 - 0.9279 = 0.0721$

c. $P(0.11 \leq \widehat{P} \leq 0.115) = P(-0.49 \leq Z \leq -0.24)$
$= P(Z \leq -0.24) - P(Z \leq -0.49)$
$= 0.4052 - 0.3121 = 0.0931$

d. Claim: $p = 0.12 \Rightarrow \widehat{P} \overset{\bullet}{\sim} N(0.12, 0.0004224)$
Experiment: $\widehat{p} = 0.09$
Likelihood: $P(\widehat{P} \leq 0.09) = 0.0721$
Conclusion: There is no evidence to suggest the
claim is false.

7.79 a. $\widehat{P} \overset{\bullet}{\sim} N(0.65, 0.0002275)$
$np = 650 \geq 5, \ n(1-p) = 350 \geq 5$

b. $P(\widehat{P} > 0.66) = P(Z > 0.66)$
$= 1 - P(Z \leq 0.66)$
$= 1 - 0.7454 = 0.2546$

c. $P(0.64 \leq \widehat{P} \leq 0.67) = P(-0.66 \leq Z \leq 1.33)$
$= P(Z \leq 1.33) - P(Z \leq -0.66)$
$= 0.9082 - 0.2546 = 0.6536$

d. $P(\widehat{P} < h) = 0.01 \Rightarrow P\left(Z < \frac{h-0.65}{\sqrt{0.0002275}}\right) = 0.01$

$\Rightarrow \frac{h-0.65}{\sqrt{0.0002275}} = -2.3263 \Rightarrow h = 0.6149$

7.81 $\widehat{P} \overset{\bullet}{\sim} N(0.31, 0.0008556)$

a. $P(0.20 \leq \widehat{P} \leq 0.30) = P(-3.76 \leq Z \leq -0.34)$
$= P(Z \leq -0.34) - P(Z \leq -3.76)$
$= 0.3669 - 0.0001 = 0.3668$

b. $P(\widehat{P} > 0.35) = P(Z > 1.37)$
$= 1 - P(Z \leq 1.37)$
$= 1 - 0.9147 = 0.0853$

c. $P(\widehat{P} < 0.28) = 0.05$

$\Rightarrow P\left(Z < \frac{-0.03}{\sqrt{(0.31)(0.69)/n}}\right) = 0.05$

$\Rightarrow \frac{-0.03}{\sqrt{(0.31)(0.69)/n}} = -1.6449 \Rightarrow n = 644$

7.83 a. $\mu = 40 \Rightarrow \overline{X} \sim N(40, 0.4^2/25)$
$2P(\overline{X} < 39.8) = 2P(Z < -2.50)$
$= 2(0.0062) = 0.0124$

b. $\mu = 40.4 \Rightarrow \overline{X} \sim N(40.4, 0.4^2/25)$
$P(39.8 \leq \overline{X} \leq 40.2) = P(-7.50 \leq Z \leq -2.50)$
$= P(Z \leq -2.50) - P(Z \leq -7.50)$
$= 0.0062 - 0.0000 = 0.0062$

c. $\mu = 39.4 \Rightarrow \overline{X} \sim N(39.4, 0.4^2/25)$
$P(39.8 \leq \overline{X} \leq 40.2) = P(5.00 \leq Z \leq 10.00)$
$\approx 1 \ (0.0000002871)$

Chapter 8: Confidence Intervals Based on a Single Sample

Section 8.1: Point Estimation

8.1 $\widehat{\theta}_2$ is the best statistic to estimate θ because it is unbiased and has small variance.

8.3 The value of the unbiased estimator is, on average, θ.

8.5 $\widehat{p} = 975/1200 = 0.8125$

8.7 $\widehat{p} = 90/500 = 0.18$

8.9 a. $x_{\min} = 95$ **b.** $x_{\max} = 104$ **c.** Interval estimate for the pressure developed: $(95, 104)$

8.11 a. $\mu = \frac{1}{40}(0.04 + \cdots + 1.44) = 0.7535$

b. $s^2 = \frac{1}{39}(30.7134 - \frac{1}{40}(908.4196)) = 0.2052$

c. $Q_1 = 0.30$, $Q_3 = 1.17$

Section 8.2: A Confidence Interval for a Population Mean When σ Is Known

8.13 a. $\bar{x} \pm z_{0.025} \frac{\sigma}{\sqrt{n}} = 15.6 \pm (1.96) \frac{3.7}{\sqrt{12}}$
$= (13.507, 17.693)$

b. $\bar{x} \pm z_{0.05} \frac{\sigma}{\sqrt{n}} = 6322 \pm (1.6449) \frac{225}{\sqrt{17}}$
$= (6232.24, 6411.76)$

c. $\bar{x} \pm z_{0.10} \frac{\sigma}{\sqrt{n}} = -45.78 \pm (1.2816) \frac{12.35}{\sqrt{9}}$
$= (-51.06, -40.50)$

d. $\bar{x} \pm z_{0.005} \frac{\sigma}{\sqrt{n}} = 0.0795 \pm (2.5758) \frac{0.006}{\sqrt{24}}$
$= (0.0763, 0.0827)$

e. $\bar{x} \pm z_{0.0005} \frac{\sigma}{\sqrt{n}} = 37.68 \pm (3.2905) \frac{2.2}{\sqrt{27}}$
$= (36.287, 39.073)$

8.15 a. $\bar{x} = \frac{1}{2}(8.55 + 10.85) = 9.7$

b. 95% CI: $(8.55, 10.85)$; 99.9% CI: $(8.40, 11.0)$. For a higher confidence level (all else being equal), the CI has to be larger.

8.17 $\bar{x} \pm z_{0.025} \frac{\sigma}{\sqrt{n}} = 106.14 \pm (1.96) \frac{35}{\sqrt{45}}$
$= (95.9139, 116.3660)$

8.19 a. $\bar{x} \pm z_{0.025} \frac{\sigma}{\sqrt{n}} = 1.4825 \pm (1.96) \frac{0.34}{\sqrt{40}}$
$= (1.3771, 1.5879)$

b. $\bar{x} \pm z_{0.005} \frac{\sigma}{\sqrt{n}} = 1.4825 \pm (2.5758) \frac{0.34}{\sqrt{40}}$
$= (1.344, 1.621)$

c. The confidence level in part (b) is larger. Therefore, the critical value is larger and the CI is larger. Intuitively, more confidence means the CI should be larger.

8.21 a. $\bar{x} \pm z_{0.005} \frac{\sigma}{\sqrt{n}} = 260 \pm (2.5758) \frac{25}{\sqrt{49}}$
$= (250.8, 269.2)$

b. There is evidence to suggest that the mean depth of icebergs is more than 200 because 200 is not included in the CI found in part (a).

8.23 a. $\bar{x} \pm z_{0.005} \frac{\sigma}{\sqrt{n}} = 20.75 \pm (2.5758) \frac{3.05}{\sqrt{35}}$
$= (19.422, 22.078)$

b. There is evidence to suggest that the true mean is less than 23. The confidence interval does not include 23 and is completely less than 23.

8.25 a. $\bar{x} \pm z_{0.05} \frac{\sigma}{\sqrt{n}} = 47500 \pm (1.6449) \frac{8500}{\sqrt{18}}$
$= (44204.49, 50795.51)$

b. $n = \left[\frac{(8500)(1.6449)}{3000} \right]^2 = 21.72 \Rightarrow n \geq 22$

c. $n = \left[\frac{(8500)(1.6449)}{1000} \right]^2 = 195.49 \Rightarrow n \geq 196$

8.27 a. $\bar{x} \pm z_{0.025} \frac{\sigma}{\sqrt{n}} = 0.131 \pm (1.96) \frac{0.02}{\sqrt{40}}$
$= (0.1248, 0.1372)$

b. It is close, but $1/8 = 0.125$ is captured by the CI in part (a). Therefore, there is no evidence to suggest the true mean is greater than 0.125. The town should not embark on the safety program.

8.29 a. $\bar{x} \pm z_{0.025} \frac{\sigma}{\sqrt{n}} = 6.6 \pm (1.96) \frac{0.5}{\sqrt{45}}$
$= (6.4539, 6.7461)$

b. $\bar{x} \pm z_{0.025} \frac{\sigma}{\sqrt{n}} = 6.6 \pm (1.96) \frac{0.25}{\sqrt{45}}$
$= (6.5270, 6.6730)$

c. $(6.4539, 6.7461)$ is an interval in which we are 95% confident the true mean wingspan lies $(\sigma = 0.5)$.

$(6.5270, 6.6730)$ is an interval in which we are 95% confident the true mean wingspan lies $(\sigma = 0.25)$.

8.31 a. $\bar{x} \pm z_{0.005} \frac{\sigma}{\sqrt{n}} = 125200 \pm (2.5758) \frac{5750}{\sqrt{30}}$
$= (122{,}495.89, 127{,}904.11)$

b. $\bar{x} \pm z_{0.005} \frac{\sigma}{\sqrt{n}} = 155900 \pm (2.5758) \frac{25390}{\sqrt{36}}$
$= (144{,}999.95, 166{,}800.05)$

c. There is evidence to suggest the mean selling price is different for the two parishes because the CIs do not overlap.

8.33 a. Football: $\bar{x} \pm z_{0.025} \frac{\sigma}{\sqrt{n}}$
$= 65.77 \pm (1.96) \frac{14.07}{\sqrt{35}} = (61.1087, 70.4313)$
Basketball: $\bar{x} \pm z_{0.025} \frac{\sigma}{\sqrt{n}}$
$= 53.90 \pm (1.96) \frac{12.5}{\sqrt{30}} = (49.4270, 58.3730)$
Hockey: $\bar{x} \pm z_{0.025} \frac{\sigma}{\sqrt{n}}$
$= 68.45 \pm (1.96) \frac{10.25}{\sqrt{32}} = (64.8986, 72.0014)$

b. There is evidence to suggest the mean coping skills level is different for football and basketball players. The CIs do not overlap.

c. Football:

$$n = \left[\frac{(14.07)(1.96)}{2}\right]^2 = 190.12 \Rightarrow n \geq 191$$

Basketball:

$$n = \left[\frac{(12.5)(1.96)}{2}\right]^2 = 150.06 \Rightarrow n \geq 151$$

Hockey:

$$n = \left[\frac{(10.25)(1.96)}{2}\right]^2 = 100.90 \Rightarrow n \geq 101$$

8.35 a. Cashews: $\bar{x} \pm z_{0.025}\frac{\sigma}{\sqrt{n}}$

$$= 5.17 \pm (1.96)\frac{0.4}{\sqrt{50}} = (5.0591, 5.2809)$$

Filberts: $\bar{x} \pm z_{0.025}\frac{\sigma}{\sqrt{n}}$

$$= 4.24 \pm (1.96)\frac{0.6}{\sqrt{50}} = (4.0737, 4.4063)$$

Pecans: $\bar{x} \pm z_{0.025}\frac{\sigma}{\sqrt{n}}$

$$= 2.6 \pm (1.96)\frac{0.95}{\sqrt{50}} = (2.3367, 2.8633)$$

There is evidence to suggest the mean amount of protein is different for cashews and pecans because the CIs do not overlap.

There is evidence to suggest the mean amount of protein is different for filberts and pecans because the CIs do not overlap.

b. Cashews: $\bar{x} \pm z_{0.025}\frac{\sigma}{\sqrt{n}}$

$$= 5.17 \pm (1.96)\frac{0.4}{\sqrt{18}} = (4.9852, 5.3548)$$

Filberts: $\bar{x} \pm z_{0.025}\frac{\sigma}{\sqrt{n}}$

$$= 4.24 \pm (1.96)\frac{0.6}{\sqrt{18}} = (3.9628, 4.5172)$$

Pecans: $\bar{x} \pm z_{0.025}\frac{\sigma}{\sqrt{n}}$

$$= 2.6 \pm (1.96)\frac{0.95}{\sqrt{18}} = (2.1611, 3.0389)$$

There is evidence to suggest the mean amount of protein is different for cashews and pecans because the CIs do not overlap.

There is evidence to suggest the mean amount of protein is different for filberts and pecans because the CIs do not overlap.

8.37 This is the *best* CI because it is the *shortest* $100(1-\alpha)\%$ CI for μ.

Section 8.3: A Confidence Interval for a Population Mean When σ Is Unknown

8.39 a. $t_{0.10} = 1.4759$ **b.** $t_{0.20} = 0.8569$
c. $t_{0.005} = 2.8609$ **d.** $t_{0.025} = 2.3646$
e. $t_{0.005} = 2.9467$ **f.** $t_{0.001} = 5.2076$
g. $t_{0.0005} = 3.7676$ **h.** $t_{0.0001} = 22.2037$

8.41 a. $t_{0.15} = 1.0931$ **b.** $t_{0.07} = 1.5286$
c. $t_{0.0025} = 3.1534$ **d.** $t_{0.01} = 2.4377$
e. $t_{0.005} = 2.6981$ **f.** $t_{0.025} = 1.9921$
g. $t_{0.02} = 2.1150$ **h.** $t_{0.003} = 2.8652$

8.43 a. $\bar{x} \pm t_{0.025}\frac{s}{\sqrt{n}} = 0.234 \pm (2.1314)\frac{0.081}{\sqrt{16}}$
$$= (0.1908, 0.2772)$$

b. $\bar{x} \pm t_{0.005}\frac{s}{\sqrt{n}} = 259.6 \pm (2.7874)\frac{76.9}{\sqrt{26}}$
$$= (217.5618, 301.6382)$$

c. $\bar{x} \pm t_{0.005}\frac{s}{\sqrt{n}} = 22.85 \pm (2.7787)\frac{7.19}{\sqrt{27}}$
$$= (19.005, 26.695)$$

d. $\bar{x} \pm t_{0.025}\frac{s}{\sqrt{n}} = 380.9 \pm (2.0860)\frac{28.4}{\sqrt{21}}$
$$= (367.9725, 393.8275)$$

e. $\bar{x} \pm t_{0.0005}\frac{s}{\sqrt{n}} = 88.1 \pm (3.9216)\frac{17.45}{\sqrt{19}}$
$$= (72.4005, 103.7995)$$

8.45 a. $\bar{x} \pm t_{0.005}\frac{s}{\sqrt{n}} = 0.2654 \pm (3.0123)\frac{0.0384}{\sqrt{13}}$
$$= (0.2345, 0.2963)$$

b. There is no evidence to suggest the true mean stell section thickness is more than 0.25 inches. The CI includes 0.25.

c. Frequency histogram:

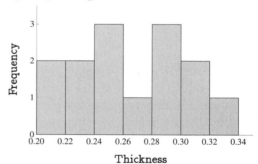

The distribution may be slightly skewed to the right. However, there are only 14 observations.

Backward Empirical Rule:
$\bar{x} = 0.2654; \ s = 0.0384$

Interval	Proportion
(0.2270, 0.3038)	0.64
(0.1886, 0.3422)	1.00
(0.1502, 0.3806)	1.00

The actual proportions are close the those given by the Empirical Rule.

$IQR/s = 0.0620/0.0384 = 1.6146$
The ratio is not close to 1.3.

Normal probability plot:

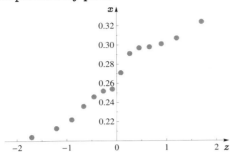

There is a slight curve to the points in this scatter plot.

There is some evidence to suggest the data are from a nonnormal population.

8.47 a. $\bar{x} \pm t_{0.025}\frac{s}{\sqrt{n}} = 1003.9167 \pm (2.2010)\frac{62.7976}{\sqrt{12}}$
$= (964.0170, 1043.8164)$

b. There is no evidence to suggest the true mean weith of adult manatees is less than 1000 pounds. The CI includes 1000.

8.49 a. $\bar{x} \pm t_{0.025}\frac{s}{\sqrt{n}} = 16.7 \pm (2.0518)\frac{3.4}{\sqrt{28}}$
$= (15.3816, 18.0184)$

b. There is no evidence to suggest the true mean route length is over 20 miles. The CI is less than 20.

8.51 a. $\bar{x} \pm t_{0.025}\frac{s}{\sqrt{n}} = 29.5241 \pm (2.1199)\frac{0.5186}{\sqrt{17}}$
$= (29.2575, 29.7908)$

b. There is evidence to suggest the true mean amount of ink in each black cartridge is under 30 ml. The CI does not include 30 and is less than 30.

8.53 a. $\bar{x} \pm t_{0.005}\frac{s}{\sqrt{n}} = 860.75 \pm (2.7969)\frac{350.50}{\sqrt{25}}$
$= (664.68, 1056.82)$

b. There is evidence to suggest the true mean property damage is greater than 500. The CI does not include 500 and is greater than 500.

8.55 $\bar{x} \pm t_{0.005}\frac{s}{\sqrt{n}} = 13.4691 \pm (2.6923)\frac{5.9809}{\sqrt{45}}$
$= (11.0687, 15.8695)$

8.57 a. $\bar{x} \pm t_{0.025}\frac{s}{\sqrt{n}} = 70.2 \pm (2.1604)\frac{2.75}{\sqrt{14}}$
$= (68.6122, 71.7878)$

b. $\bar{x} \pm t_{0.025}\frac{s}{\sqrt{n}} = 72.1 \pm (2.2281)\frac{1.55}{\sqrt{11}}$
$= (71.0587, 73.1413)$

c. It is assumed that the underlying populations are normal.

d. There is no evidence to suggest the New England mean thermostat setting is different from the Southern thermostat setting during winter because the CIs overlap.

8.59 a. $\bar{x} \pm t_{0.01}\frac{s}{\sqrt{n}} = 285.33 \pm (2.5669)\frac{5.9607}{\sqrt{18}}$
$= (281.7269, 288.9397)$

b. $\bar{x} \pm t_{0.01}\frac{s}{\sqrt{n}} = 257.75 \pm (2.4999)\frac{16.3581}{\sqrt{24}}$
$= (249.4027, 266.0973)$

c. There is evidence to suggest the true mean beats per minute in atrial flutter is different for men and women because the CIs do not overlap.

d. The normality assumption seems reasonable. Atrial flutter is a measurement, and the distribution is probably not skewed in either direction.

8.61 a. $\bar{x} \pm t_{0.005}\frac{s}{\sqrt{n}} = 12.9985 \pm (2.7787)\frac{2.4317}{\sqrt{27}}$
$= (11.6981, 14.2989)$

b. $\bar{x} \pm t_{0.005}\frac{s}{\sqrt{n}} = 8.7536 \pm (2.7969)\frac{1.6997}{\sqrt{25}}$
$= (7.8028, 9.7044)$

c. Rural areas:
Frequency histogram:

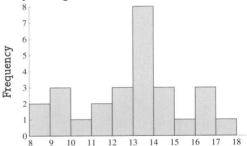

Response time (rural)

The shape of the distribution is approximately normal.

Backward Empirical Rule:
$\bar{x} = 12.9985$, $s = 2.4317$

Interval	Proportion
$(10.5668, 15.4302)$	0.67
$(8.1351, 17.8619)$	1.00
$(5.7035, 20.2936)$	1.00

The actual proportions are fairly close the those given by the Empirical Rule.

$IQR/s = 2.95/2.4317 = 1.2131$
The ratio is fairly close to 1.3.

Normal probability plot:

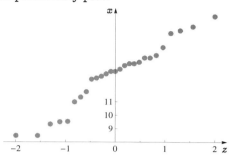

There is some evidence of a curve in this plot towards the bottom left.

There is insufficient evidence to suggest the data are from a nonnormal population.

City areas:
Frequency histogram:

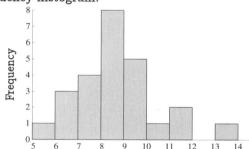

Response time (cities)

The shape of the distribution is approximately normal.

Backward Empirical Rule:
$\bar{x} = 8.7536$, $s = 1.6997$

Interval	Proportion
$(7.0539, 10.4533)$	0.72
$(5.3543, 12.1529)$	0.96
$(3.6546, 13.8526)$	1.00

The actual proportions are close the those given by the Empirical Rule.

$IQR/s = 1.96/1.6997 = 1.1531$
The ratio is fairly close to 1.3.

Normal probability plot:

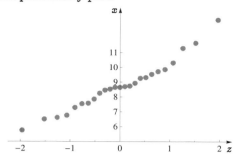

The points lie along a fairly straight line.

There is insufficient evidence to suggest the data are from a nonnormal population.

d. There is evidence to suggest the true mean response time is different for rural and city areas since the CIs do not overlap.

8.63 a. Men: $\bar{x} \pm t_{0.025} \frac{s}{\sqrt{n}}$

$= 38.9 \pm (2.0010) \frac{7.9}{\sqrt{60}} = (36.86, 40.94)$

Women: $\bar{x} \pm t_{0.025} \frac{s}{\sqrt{n}}$

$= 35.6 \pm (2.0227) \frac{4.5}{\sqrt{40}} = (34.16, 37.04)$

b. There is no evidence to suggest the mean age for men is different from the mean age for women. The CIs overlap.

c. $\bar{x} \pm t_{0.005} \frac{s}{\sqrt{n}} = 257.5 \pm (2.6439) \frac{56.8}{\sqrt{75}}$

$= (240.1594, 274.8406)$

This is an interval in which we are 99% confident the true mean distance traveled lies.

Section 8.4: A Large-Sample Confidence Interval for a Population Proportion

8.65 a. $\hat{p} = 85/105 = 0.8095$
$(105)(0.8095) = 85 \geq 5$, $(105)(0.1905) = 20 \geq 5$
$\widehat{P}$ is approximately normal.

b. $\hat{p} = 1645/1750 = 0.94$
$(1750)(0.94) = 1645 \geq 5$, $(1750)(0.06) = 105 \geq 5$
$\widehat{P}$ is approximately normal.

c. $\hat{p} = 220/225 = 0.9778$
$(225)(0.9778) = 220 \geq 5$, $(225)(0.0222) = 5 \geq 5$
$\widehat{P}$ is approximately normal (just barely).

d. $\hat{p} = 3/183 = 0.0164$
$(183)(0.0164) = 3 < 5$, $(183)(0.9836) = 180 \geq 5$
$\widehat{P}$ is not approximately normal.

e. $\hat{p} = 350/377 = 0.9284$
$(377)(0.9284) = 350 \geq 5$, $(377)(0.0716) = 27 \geq 5$
$\widehat{P}$ is approximately normal.

f. $\widehat{p} = 478/480 = 0.9958$

$(480)(0.9958) = 478 \geq 5$, $(480)(0.0042) = 2 < 5$

$\widehat{P}$ is not approximately normal.

8.67 a. $\widehat{p} \pm z_{0.005}\sqrt{\dfrac{\widehat{p}(1-\widehat{p})}{n}}$

$= 0.7493 \pm (2.5758)\sqrt{\dfrac{(0.7493)(0.2507)}{1336}}$

$= (0.7187, 0.7798)$

b. $\widehat{p} \pm z_{0.025}\sqrt{\dfrac{\widehat{p}(1-\widehat{p})}{n}}$

$= 0.8774 \pm (1.96)\sqrt{\dfrac{(0.8774)(0.1226)}{775}}$

$= (0.8543, 0.9005)$

c. $\widehat{p} \pm z_{0.025}\sqrt{\dfrac{\widehat{p}(1-\widehat{p})}{n}}$

$= 0.4824 \pm (1.96)\sqrt{\dfrac{(0.4824)(0.5176)}{85}}$

$= (0.3761, 0.5886)$

d. $\widehat{p} \pm z_{0.01}\sqrt{\dfrac{\widehat{p}(1-\widehat{p})}{n}}$

$= 0.8657 \pm (2.3263)\sqrt{\dfrac{(0.8657)(0.1343)}{335}}$

$= (0.8223, 0.9090)$

e. $\widehat{p} \pm z_{0.0005}\sqrt{\dfrac{\widehat{p}(1-\widehat{p})}{n}}$

$= 0.0830 \pm (3.2905)\sqrt{\dfrac{(0.0830)(0.9170)}{566}}$

$= (0.0449, 0.1212)$

8.69 a. $n = (0.5)(0.5)\left[\dfrac{2.5758}{0.06}\right]^2$

$= 460.76 \Rightarrow n \geq 461$

b. $n = (0.5)(0.5)\left[\dfrac{1.96}{0.10}\right]^2 = 96.04 \Rightarrow n \geq 97$

c. $n = (0.5)(0.5)\left[\dfrac{2.3263}{0.002}\right]^2$

$= 338{,}243.4 \Rightarrow n \geq 338{,}244$

d. $n = (0.5)(0.5)\left[\dfrac{1.6449}{0.05}\right]^2 = 270.55 \Rightarrow n \geq 271$

e. $n = (0.5)(0.5)\left[\dfrac{3.2905}{0.2}\right]^2 = 67.67 \Rightarrow n \geq 68$

8.71 a. As B becomes smaller, the fraction $z_{\alpha/2}/B$ increases. Therefore, n increases.

b. As the confidence level increases, $z_{\alpha/2}$ increases and the fraction $z_{\alpha/2}/B$ increases. Therefore, n increases.

c. As $\widehat{p}$ gets closer to zero, the product $\widehat{p}(1-\widehat{p})$ gets smaller. Therefore, n decreases.

d. As $\widehat{p}$ gets closer to 1, the product $\widehat{p}(1-\widehat{p})$ gets smaller. Therefore, n decreases.

8.73 a. $\widehat{p} \pm z_{0.05}\sqrt{\dfrac{\widehat{p}(1-\widehat{p})}{n}}$

$= 0.4087 \pm (1.6449)\sqrt{\dfrac{(0.4087)(0.5913)}{575}}$

$= (0.3750, 0.4424)$

b. $n = (0.4)(0.6)\left[\dfrac{1.96}{0.03}\right]^2 = 1024.39 \Rightarrow n \geq 1025$

8.75 a. $\widehat{p} \pm z_{0.01}\sqrt{\dfrac{\widehat{p}(1-\widehat{p})}{n}}$

$= 0.12 \pm (2.3263)\sqrt{\dfrac{(0.12)(0.88)}{125}}$

$= (0.0524, 0.1876)$

b. $n = (0.5)(0.5)\left[\dfrac{2.3263}{0.05}\right]^2 = 541.19 \Rightarrow n \geq 542$

8.77 a. $\widehat{p} \pm z_{0.025}\sqrt{\dfrac{\widehat{p}(1-\widehat{p})}{n}}$

$= 0.4403 \pm (1.96)\sqrt{\dfrac{(0.4403)(0.5597)}{2126}}$

$= (0.4192, 0.4614)$

b. $\widehat{p} \pm z_{0.025}\sqrt{\dfrac{\widehat{p}(1-\widehat{p})}{n}}$

$= 0.3100 \pm (1.96)\sqrt{\dfrac{(0.3100)(0.6900)}{2126}}$

$= (0.2903, 0.3296)$

c. $\widehat{p} \pm z_{0.025}\sqrt{\dfrac{\widehat{p}(1-\widehat{p})}{n}}$

$= 0.2399 \pm (1.96)\sqrt{\dfrac{(0.2399)(0.7601)}{2126}}$

$= (0.2217, 0.2580)$

8.79 a. $\widehat{p} \pm z_{0.025}\sqrt{\dfrac{\widehat{p}(1-\widehat{p})}{n}}$

$= 0.3245 \pm (1.96)\sqrt{\dfrac{(0.3245)(0.6755)}{188}}$

$= (0.2575, 0.3914)$

b. $n = (0.25)(0.75)\left[\dfrac{1.96}{0.025}\right]^2 = 1152.44 \Rightarrow n \geq 1153$

8.81 a. $\widehat{p} \pm z_{0.005}\sqrt{\dfrac{\widehat{p}(1-\widehat{p})}{n}}$

$= 0.3205 \pm (2.5758)\sqrt{\dfrac{(0.3205)(0.6795)}{1560}}$

$= (0.2901, 0.3509)$

b. There is no evidence to suggest the true proportion has changed since 0.30 is included in the CI (just barely).

8.83 a. Northeast: $\widehat{p} \pm z_{0.005}\sqrt{\dfrac{\widehat{p}(1-\widehat{p})}{n}}$

$= 0.80 \pm (2.5758)\sqrt{\dfrac{(0.80)(0.20)}{225}}$

$= (0.7313, 0.8687)$

Midwest: $\widehat{p} \pm z_{0.005}\sqrt{\dfrac{\widehat{p}(1-\widehat{p})}{n}}$

$= 0.8116 \pm (2.5758)\sqrt{\dfrac{(0.8116)(0.1884)}{276}}$

$= (0.7510, 0.8722)$

South Central: $\widehat{p} \pm z_{0.005}\sqrt{\dfrac{\widehat{p}(1-\widehat{p})}{n}}$

$= 0.7708 \pm (2.5758)\sqrt{\dfrac{(0.7708)(0.2292)}{301}}$

$= (0.7084, 0.8332)$

South Atlantic: $\widehat{p} \pm z_{0.005}\sqrt{\dfrac{\widehat{p}(1-\widehat{p})}{n}}$

$= 0.8304 \pm (2.5758)\sqrt{\frac{(0.8304)(0.1696)}{454}}$

$= (0.7850, 0.8758)$

West: $\widehat{p} \pm z_{0.005}\sqrt{\frac{\widehat{p}(1-\widehat{p})}{n}}$

$= 0.8306 \pm (2.5758)\sqrt{\frac{(0.8306)(0.1694)}{366}}$

$= (0.7801, 0.8811)$

b. The CI corresponding to the Northeast region is the largest. The sample size is the smallest.

8.85 a. Secondary school: $\widehat{p} \pm z_{0.005}\sqrt{\frac{\widehat{p}(1-\widehat{p})}{n}}$

$= 0.0778 \pm (2.5758)\sqrt{\frac{(0.0778)(0.9222)}{257}}$

$= (0.0348, 0.1209)$

Other post-secondary: $\widehat{p} \pm z_{0.005}\sqrt{\frac{\widehat{p}(1-\widehat{p})}{n}}$

$= 0.0711 \pm (2.5758)\sqrt{\frac{(0.0711)(0.9289)}{380}}$

$= (0.0371, 0.1050)$

Post-secondary: $\widehat{p} \pm z_{0.005}\sqrt{\frac{\widehat{p}(1-\widehat{p})}{n}}$

$= 0.0361 \pm (2.5758)\sqrt{\frac{(0.0361)(0.9639)}{305}}$

$= (0.0086, 0.0636)$

b. There is no evidence the true proportions are different for post-secondary and secondary school since the CIs overlap.

There is no evidence the true proportions are different for post-secondary and other post-secondary since the CIs overlap.

8.87 a. $\widehat{p} \pm z_{0.025}\sqrt{\frac{\widehat{p}(1-\widehat{p})}{n}}$

$= 0.36 \pm (1.96)\sqrt{\frac{(0.36)(0.64)}{481}}$

$= (0.3171, 0.4029)$

b. $\widehat{p} \pm z_{0.025}\sqrt{\frac{\widehat{p}(1-\widehat{p})}{n}}$

$= 0.40 \pm (1.96)\sqrt{\frac{(0.40)(0.60)}{1028}}$

$= (0.3701, 0.4299)$

c. There is no evidence to suggest the true proportions are different. The CIs overlap.

8.89 a. Treatment: $\widehat{p} \pm z_{0.025}\sqrt{\frac{\widehat{p}(1-\widehat{p})}{n}}$

$= 0.1312 \pm (1.96)\sqrt{\frac{(0.1312)(0.8688)}{465}}$

$= (0.1005, 0.1619)$

Placebo: $\widehat{p} \pm z_{0.025}\sqrt{\frac{\widehat{p}(1-\widehat{p})}{n}}$

$= 0.0974 \pm (1.96)\sqrt{\frac{(0.0974)(0.9026)}{154}}$

$= (0.0506, 0.1442)$

b. There is no evidence to suggest the true proportion of people who suffer from headaches in the two groups is different. The CIs overlap.

c. Treatment: $\widehat{p} \pm z_{0.01}\sqrt{\frac{\widehat{p}(1-\widehat{p})}{n}}$

$= 0.0667 \pm (2.3263)\sqrt{\frac{(0.0667)(0.9333)}{465}}$

$= (0.0398, 0.0936)$

Placebo: $\widehat{p} \pm z_{0.01}\sqrt{\frac{\widehat{p}(1-\widehat{p})}{n}}$

$= 0.0584 \pm (2.3263)\sqrt{\frac{(0.0584)(0.9416)}{154}}$

$= (0.0145, 0.1024)$

d. There is no evidence to suggest the true proportion of people who suffer from a rash in the two groups is different. The CIs overlap.

8.91 $P\left(-z_{\alpha/2} < \frac{\widehat{P}-p}{\sqrt{\frac{p(1-p)}{n}}} < z_{\alpha/2}\right) = 1 - \alpha$

$\Rightarrow P\left(\frac{\widehat{P}^2 - 2\widehat{P}p + p^2}{\frac{p(1-p)}{n}} < z_{\alpha/2}^2\right) = 1 - \alpha$

$\Rightarrow P\left(\widehat{P}^2 - 2\widehat{P}p + p^2 < z_{\alpha/2}^2\frac{p(1-p)}{n}\right) = 1 - \alpha$

$\Rightarrow P\left(\left(1 + \frac{z_{\alpha/2}^2}{n}\right)p^2 - \left(2\widehat{P} + \frac{z_{\alpha/2}^2}{n}\right)p + \widehat{P}^2 < 0\right)$

$= 1 - \alpha$

Using the quadratic formula, the endpoints are:

$\frac{n\widehat{p} + z_{\alpha/2}^2/2}{n + z_{\alpha/2}^2} \pm \frac{z_{\alpha/2}\sqrt{n}}{n + z_{\alpha/2}^2}\sqrt{\widehat{P}(1-\widehat{P}) + z_{\alpha/2}^2/(4n)}$

a. $\widehat{p} \pm z_{0.025}\sqrt{\frac{\widehat{p}(1-\widehat{p})}{n}}$

$= 0.60 \pm (1.96)\sqrt{\frac{(0.60)(0.40)}{100}}$

$= (0.5040, 0.6960)$

Using the Wilson interval;

$\frac{(100)(0.6) + 1.96^2/2}{100 + 1.96^2} \pm \frac{1.96\sqrt{100}}{100 + 1.96^2}\sqrt{(0.6)(0.4) + 1.96^2/400}$

$= (0.5020, 0.6906)$

The Wilson interval is shorter because it is more precise.

b.

n	Traditional CI	Wilson CI
120	$(0.5123, 0.6877)$	$(0.5106, 0.6832)$
140	$(0.5188, 0.6812)$	$(0.5172, 0.6774)$
160	$(0.5241, 0.6759)$	$(0.5226, 0.6727)$
180	$(0.5284, 0.6716)$	$(0.5271, 0.6688)$
200	$(0.5321, 0.6679)$	$(0.5308, 0.6654)$
220	$(0.5353, 0.6647)$	$(0.5341, 0.6625)$
240	$(0.5380, 0.6620)$	$(0.5369, 0.6599)$
260	$(0.5405, 0.6595)$	$(0.5394, 0.6577)$
280	$(0.5426, 0.6574)$	$(0.5416, 0.6557)$
300	$(0.5446, 0.6554)$	$(0.5436, 0.6538)$
320	$(0.5463, 0.6537)$	$(0.5454, 0.6522)$
340	$(0.5479, 0.6521)$	$(0.5471, 0.6507)$
360	$(0.5494, 0.6506)$	$(0.5486, 0.6493)$
380	$(0.5507, 0.6493)$	$(0.5500, 0.6480)$
400	$(0.5520, 0.6480)$	$(0.5513, 0.6468)$
420	$(0.5531, 0.6469)$	$(0.5524, 0.6457)$
440	$(0.5542, 0.6458)$	$(0.5535, 0.6447)$
460	$(0.5552, 0.6448)$	$(0.5546, 0.6438)$
480	$(0.5562, 0.6438)$	$(0.5555, 0.6429)$
500	$(0.5571, 0.6429)$	$(0.5565, 0.6420)$

As n increases in the Wilson CI, $\widehat{p}$ is closer to the center of the interval.

Section 8.5: A Confidence Interval for a Population Variance

8.93 a. $\chi^2_{0.10} = 9.2364$ **b.** $\chi^2_{0.001} = 61.0983$
c. $\chi^2_{0.05} = 26.2962$ **d.** $\chi^2_{0.025} = 35.4789$
e. $\chi^2_{0.99} = 3.0535$ **f.** $\chi^2_{0.95} = 7.2609$
g. $\chi^2_{0.975} = 11.6886$ **h.** $\chi^2_{0.995} = 1.7349$

8.95 a. df $= 21$, $\alpha/2 = 0.025$, $1 - \alpha/2 = 0.975$
$\chi^2_{0.975} = 10.2829$, $\chi^2_{0.025} = 35.4789$

b. df $= 36$, $\alpha/2 = 0.005$, $1 - \alpha/2 = 0.995$
$\chi^2_{0.995} = 17.8867$, $\chi^2_{0.005} = 61.5812$

c. df $= 10$, $\alpha/2 = 0.01$, $1 - \alpha/2 = 0.99$
$\chi^2_{0.99} = 2.5582$, $\chi^2_{0.01} = 23.2093$

d. df $= 30$, $\alpha/2 = 0.05$, $1 - \alpha/2 = 0.95$
$\chi^2_{0.95} = 18.4927$, $\chi^2_{0.05} = 43.7730$

e. df $= 4$, $\alpha/2 = 0.025$, $1 - \alpha/2 = 0.975$
$\chi^2_{0.975} = 0.4844$, $\chi^2_{0.025} = 11.1433$

f. df $= 36$, $\alpha/2 = 0.0005$, $1 - \alpha/2 = 0.9995$
$\chi^2_{0.9995} = 14.4012$, $\chi^2_{0.0005} = 70.5881$

8.97 a. $\frac{(n-1)s^2}{\chi^2_{0.005}} = \frac{(13)(3.08)}{29.8195} = 1.3427$

$\frac{(n-1)s^2}{\chi^2_{0.995}} = \frac{(13)(3.08)}{3.5650} = 11.2313$

CI for σ^2: $(1.3427, 11.2313)$
CI for σ: $(1.1588, 3.3513)$

b. $\frac{(n-1)s^2}{\chi^2_{0.025}} = \frac{(10)(64.1)}{20.4832} = 31.2940$

$\frac{(n-1)s^2}{\chi^2_{0.975}} = \frac{(10)(64.1)}{3.2470} = 197.4147$

CI for σ^2: $(31.2940, 197.4147)$
CI for σ: $(5.5941, 14.0504)$

c. $\frac{(n-1)s^2}{\chi^2_{0.10}} = \frac{(5)(59.07)}{9.2364} = 31.9769$

$\frac{(n-1)s^2}{\chi^2_{0.90}} = \frac{(5)(59.07)}{1.6103} = 183.4121$

CI for σ^2: $(31.9769, 183.4121)$
CI for σ: $(5.6548, 13.5430)$

d. $\frac{(n-1)s^2}{\chi^2_{0.0001}} = \frac{(26)(7.35)}{61.6573} = 3.0994$

$\frac{(n-1)s^2}{\chi^2_{0.9999}} = \frac{(26)(7.35)}{7.1998} = 26.5425$

CI for σ^2: $(3.0994, 26.5425)$
CI for σ: $(1.7605, 5.1519)$

e. $\frac{(n-1)s^2}{\chi^2_{0.025}} = \frac{(21)(31.38)}{35.4789} = 18.5739$

$\frac{(n-1)s^2}{\chi^2_{0.975}} = \frac{(21)(31.38)}{10.2829} = 64.0850$

CI for σ^2: $(18.5739, 64.0850)$
CI for σ: $(4.3097, 8.0053)$

f. $\frac{(n-1)s^2}{\chi^2_{0.005}} = \frac{(17)(12.39)}{35.7185} = 5.8969$

$\frac{(n-1)s^2}{\chi^2_{0.995}} = \frac{(17)(12.39)}{5.6972} = 36.9707$

CI for σ^2: $(5.8969, 36.9707)$
CI for σ: $(2.4284, 6.0804)$

8.99 a. $\frac{(n-1)s^2}{\chi^2_{0.025}} = \frac{(24)(4.25)}{39.3641} = 2.5912$

$\frac{(n-1)s^2}{\chi^2_{0.975}} = \frac{(24)(4.25)}{12.4012} = 8.2250$

CI for σ^2: $(2.5912, 8.2250)$

b. CI for σ: $(\sqrt{2.5912}, \sqrt{8.2250}) = (1.6097, 2.8679)$

8.101 a. $\frac{(n-1)s^2}{\chi^2_{0.01}} = \frac{(17)(3.3859)}{33.4087} = 1.7229$

$\frac{(n-1)s^2}{\chi^2_{0.99}} = \frac{(17)(3.3859)}{6.4078} = 8.9829$

CI for σ^2: $(1.7229, 8.9829)$

b. Frequency histogram:

The shape of the distribution does not appear to be normal.

Backward Empirical Rule:
$\bar{x} = 2.8294$; $s = 1.8401$

Interval	Proportion
(0.9894, 4.6695)	0.67
(−0.8507, 6.5096)	0.94
(−2.6908, 8.3497)	1.00

The actual proportions are close the those given by the Empirical Rule.

$IQR/s = 2.42/1.84 = 1.3152$
The ratio is very close to 1.3.

Normal probability plot:

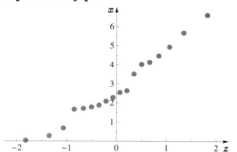

There is a slight curve to the points in this scatter plot.

There is no overwhelming evidence to suggest the data are from a non-normal population.

8.103 a. $\frac{(n-1)s^2}{\chi^2_{0.025}} = \frac{(29)(5.4989)}{45.7223} = 3.4877$

$\frac{(n-1)s^2}{\chi^2_{0.975}} = \frac{(29)(5.4989)}{16.0471} = 9.9374$

CI for σ^2: $(3.4877, 9.9374)$

b. Frequency histogram:

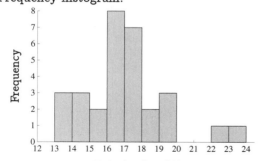

Relative humidity

The shape of the distribution is approximately normal.

Backward Empirical Rule:
$\bar{x} = 16.5333$; $s = 2.3450$

Interval	Proportion
(14.1884, 18.8783)	0.63
(11.8434, 21.2233)	0.93
(9.4984, 23.5682)	1.00

The actual proportions are close the those given by the Empirical Rule.

$IQR/s = 2.0/2.3450 = 0.853$
The ratio is not very close to 1.3.

Normal probability plot:

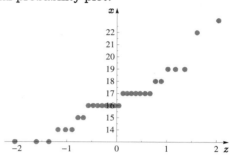

These points do not appear to lie along a straight line.

There is some evidence to suggest the data are from a nonnormal population. IQR/s is far away from 1.3, and the normal probability plot is not very linear.

8.105 a. $\frac{(n-1)s^2}{\chi^2_{0.05}} = \frac{(16)(0.55)}{26.2962} = 0.3346$

$\frac{(n-1)s^2}{\chi^2_{0.95}} = \frac{(16)(0.55)}{7.9616} = 1.1053$

CI for σ^2: $(0.3346, 1.1053)$

b. CI for σ: $(\sqrt{0.3346}, \sqrt{1.1053}) = (0.5785, 1.0513)$

c. There is no evidence to suggest the standard deviation is greater than 1. The CI for σ includes 1.

8.107 a. $\frac{(n-1)s^2}{\chi^2_{0.005}} = \frac{(29)(1.0126)}{52.3356} = 0.5611$

$\frac{(n-1)s^2}{\chi^2_{0.995}} = \frac{(29)(1.0126)}{13.1211} = 2.2381$

CI for σ_V^2: $(0.5611, 2.2381)$

b. $\frac{(n-1)s^2}{\chi^2_{0.005}} = \frac{(24)(29.5933)}{45.5585} = 15.5896$

$\frac{(n-1)s^2}{\chi^2_{0.995}} = \frac{(24)(29.5933)}{9.8862} = 71.8413$

CI for σ_R^2: $(15.5896, 71.8413)$

c. The veteran will probably get the starting job. The CI suggests the population variance for the veteran is much smaller than for the rookie.

8.109 a. $\frac{(n-1)s^2}{\chi^2_{0.01}} = \frac{(19)(0.0608)}{36.1909} = 0.0319$

$\frac{(n-1)s^2}{\chi^2_{0.99}} = \frac{(19)(0.0608)}{7.6327} = 0.1515$

CI for σ^2: $(0.0319, 0.1515)$

b. CI for σ: $(\sqrt{0.0319}, \sqrt{0.1515}) = (0.1787, 0.3892)$

c. Frequency histogram:

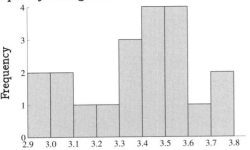

Water table depth

The shape of the distribution does not appear to be normal.

Backward Empirical Rule:

$\overline{x} = 3.371$; $s = 0.2467$

Interval	Proportion
$(3.1243, 3.6177)$	0.65
$(2.8777, 3.8643)$	1.00
$(2.6310, 4.1110)$	1.00

The actual proportions are close the those given by the Empirical Rule.

$IQR/s = 0.31/0.2467 = 1.257$

The ratio is very close to 1.3.

Normal probability plot:

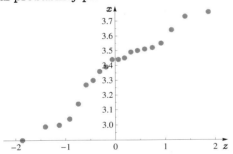

These points do not appear to lie along a straight line.

There is some evidence to suggest the data are from a nonnormal population. The histogram does not appear normal, and the normal probability plot has distinct curves.

8.111 a.

$\frac{(n-1)s^2}{\chi^2_{0.025}} = \frac{(14)(2.51502 \times 10^{16})}{26.1189} = 1.3481 \times 10^{16}$

$\frac{(n-1)s^2}{\chi^2_{0.975}} = \frac{(14)(2.51502 \times 10^{16})}{5.6287} = 6.2555 \times 10^{16}$

CI for σ^2: $(1.3481 \times 10^{16}, 6.2555 \times 10^{16})$

b. $\frac{(n-1)s^2}{\chi^2_{0.025}} = \frac{(14)(1.2194)}{26.1189} = 0.6536$

$\frac{(n-1)s^2}{\chi^2_{0.975}} = \frac{(14)(1.2194)}{5.6287} = 3.0330$

CI for σ^2_r: $(0.6536, 3.0330)$

c. There is no evidence to suggest the population variance is greater than 1. The CI includes 1.

8.113 a. $\frac{(n-1)s^2}{\chi^2_{0.025}} = \frac{(20)(11.1911)}{34.1696} = 6.5503$

$\frac{(n-1)s^2}{\chi^2_{0.975}} = \frac{(20)(11.1911)}{9.5908} = 23.3373$

CI for σ^2: $(6.5503, 23.3373)$

b. There is no evidence to suggest the population variance is less than 12. The CI includes 12.

8.115 a. $\frac{(n-1)s^2}{\chi^2_{0.025}} = \frac{(29)(1.5)}{45.7223} = 0.9514$

$\frac{(n-1)s^2}{\chi^2_{0.975}} = \frac{(29)(1.5)}{16.0471} = 2.7108$

CI for σ^2_N: $(0.9514, 2.7108)$

b. $\frac{(n-1)s^2}{\chi^2_{0.025}} = \frac{(21)(0.89)}{35.4789} = 0.5268$

$\frac{(n-1)s^2}{\chi^2_{0.975}} = \frac{(21)(0.89)}{10.2829} = 1.8176$

CI for σ^2_G: $(0.5268, 1.8176)$

c. There is no evidence to suggest that the population variance in height is different for natural rice plants and genetically engineered rice plants. The CIs overlap.

8.117 a. Cirrus clouds:

$\frac{(n-1)s^2}{\chi^2_{0.005}} = \frac{(10)(0.06)}{25.1882} = 0.0238$

$\frac{(n-1)s^2}{\chi^2_{0.995}} = \frac{(10)(0.06)}{2.1559} = 0.2783$

CI for σ^2_V: $(0.0238, 0.2783)$

$\frac{(n-1)s^2}{\chi^2_{0.995}} = \frac{(16)(201.7)}{34.2672} = 94.1776$

$\frac{(n-1)s^2}{\chi^2_{0.995}} = \frac{(16)(201.7)}{5.1422} = 627.5906$

CI for σ^2_{IB}: $(94.1776, 627.5906)$

b. Cumulus clouds:

$\frac{(n-1)s^2}{\chi^2_{0.005}} = \frac{(20)(0.08)}{39.9968} = 0.0400$

$\frac{(n-1)s^2}{\chi^2_{0.995}} = \frac{(20)(0.08)}{7.4338} = 0.2152$

CI for σ^2_V: $(0.0400, 0.2152)$

$\frac{(n-1)s^2}{\chi^2_{0.005}} = \frac{(27)(225.6)}{49.6449} = 122.6953$

$\frac{(n-1)s^2}{\chi^2_{0.995}} = \frac{(27)(225.6)}{11.8076} = 515.8717$

CI for σ^2_{IB}: $(122.6953, 515.8717)$

c. There is no evidence to suggest that the population variance in column water vapor is different for cirrus and cumulus clouds. The CIs overlap.

There is no evidence to suggest that the population

variance in IB temperature is different for cirrus and cumulus clouds. The CIs overlap.

8.119 a. Bounded above:

$$P\left(\frac{(n-1)S^2}{\sigma^2} > \chi^2_{1-\alpha}\right) = 1 - \alpha$$

$$\Rightarrow P\left(\frac{(n-1)S^2}{\chi^2_{1-\alpha}} > \sigma^2\right) = 1 - \alpha$$

CI for σ^2 bounded above: $0 < \sigma^2 < \frac{(n-1)s^2}{\chi^2_{1-\alpha}}$

Bounded below:

$$P\left(\frac{(n-1)S^2}{\sigma^2} < \chi^2_{\alpha}\right) = 1 - \alpha$$

$$\Rightarrow P\left(\frac{(n-1)S^2}{\chi^2_{\alpha}} < \sigma^2\right) = 1 - \alpha \text{ CI for } \sigma^2 \text{ bounded}$$

below: $\sigma^2 > \frac{(n-1)s^2}{\chi^2_{\alpha}}$

b. $\frac{(n-1)s^2}{\chi^2_{0.05}} = \frac{(23)(501,247,481.8)}{13.0905} = 880,690,545.5$

CI for σ^2: $0 < \sigma^2 < 880,690,545.5$

Chapter Exercises

8.121 a. $\bar{x} \pm t_{0.025}\frac{s}{\sqrt{n}} = 36.2208 \pm (2.0687)\frac{9.2514}{\sqrt{24}}$

$= (32.3143, 40.1274)$

b. $\frac{(n-1)s^2}{\chi^2_{0.025}} = \frac{(23)(85.5887)}{38.0756} = 51.7008$

$\frac{(n-1)s^2}{\chi^2_{0.975}} = \frac{(23)(85.5887)}{11.6886} = 168.4160$

CI for σ^2: $(51.7008, 168.4160)$

c. Normal probability plot:

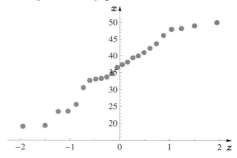

The points do not appear to fall along a straight line. There is some evidence to suggest the data are from a nonnormal population.

8.123 a. $\hat{p} = 189/270 = 0.70$

$(270)(0.70) = 189 \geq 5,\ (270)(0.30) = 81 \geq 5$

b. $\hat{p} \pm z_{0.005}\sqrt{\frac{\hat{p}(1-\hat{p})}{n}}$

$= 0.70 \pm (2.5758)\sqrt{\frac{(0.70)(0.30)}{270}}$

$= (0.6282, 0.7718)$

c. $n = (0.5)(0.5)\left[\frac{2.5758}{0.05}\right]^2 = 663.49 \Rightarrow n \geq 664$

8.125 a. $\hat{p} = 280/500 = 0.56$

$(500)(0.56) = 280 \geq 5,\ (500)(0.44) = 220 \geq 5$

b. $\hat{p} \pm z_{0.005}\sqrt{\frac{\hat{p}(1-\hat{p})}{n}}$

$= 0.56 \pm (2.5758)\sqrt{\frac{(0.56)(0.44)}{500}}$

$= (0.5028, 0.6172)$

c. There is no evidence to refute the claim. The CI for p includes 0.60.

8.127 a. $\hat{p} \pm z_{0.05}\sqrt{\frac{\hat{p}(1-\hat{p})}{n}}$

$= 0.0025 \pm (1.6449)\sqrt{\frac{(0.0025)(0.9975)}{8755}}$

$= (0.0016, 0.0034)$

b. There is no evidence to suggest the proportion is greater than 0.5%. The CI in for p is less than 0.005.

8.129 a. $\bar{x} \pm t_{0.005}\frac{s}{\sqrt{n}} = 16.1286 \pm (2.7284)\frac{2.7089}{\sqrt{35}}$

$= (14.8793, 17.3779)$

b. $\frac{(n-1)s^2}{\chi^2_{0.005}} = \frac{(34)(7.3380)}{58.9639} = 4.2314$

$\frac{(n-1)s^2}{\chi^2_{0.995}} = \frac{(34)(7.3380)}{16.5013} = 15.1195$

CI for σ^2: $(4.2313, 15.1195)$

c. We assumed the underlying population is normal.

d. There is no evidence to suggest the mean sweetener consumed per day is greater than 15. The CI for the population mean includes 15.

8.131 a. $\bar{x} \pm t_{0.005}\frac{s}{\sqrt{n}} = 128.00 \pm (2.9208)\frac{75.5}{\sqrt{17}}$

$= (74.52, 181.48)$

b. There is no evidence to suggest the mean amount of merchandise lost to shoplifters is greater than \$100. The CI includes 100.

c. $\frac{(n-1)s^2}{\chi^2_{0.005}} = \frac{(16)(5700.25)}{34.2672} = 1662.56$

$\frac{(n-1)s^2}{\chi^2_{0.995}} = \frac{(16)(5700.25)}{5.1422} = 17736.36$

CI for σ^2: $(2661.56, 17736.36)$

d. There is evidence to suggest the population variance is greater than 2500. The CI does not include 2500 and is greater than 2500.

8.133 a. Caucasian: $\hat{p} \pm z_{0.025}\sqrt{\frac{\hat{p}(1-\hat{p})}{n}}$

$= 0.1361 \pm (1.96)\sqrt{\frac{(0.1361)(0.8639)}{1220}}$

$= (0.1168, 0.1553)$

African American: $\hat{p} \pm z_{0.025}\sqrt{\frac{\hat{p}(1-\hat{p})}{n}}$

$= 0.1898 \pm (1.96)\sqrt{\frac{(0.1898)(0.8102)}{1080}}$

$= (0.1664, 0.2132)$

Hispanic: $\hat{p} \pm z_{0.025}\sqrt{\frac{\hat{p}(1-\hat{p})}{n}}$

$= 0.3322 \pm (1.96)\sqrt{\frac{(0.3322)(0.6678)}{1156}}$

$= (0.3050, 0.3593)$

b. There is evidence to suggest the proportion of African Americans and the proportion of Hispanics without health insurance is different from 0.146. The two CIs do not include 0.146.

8.135 a. $\bar{x} \pm t_{0.025} \frac{s}{\sqrt{n}} = 0.30 \pm (2.0117) \frac{0.058}{\sqrt{48}}$

$= (0.2832, 0.3168)$

b. $\bar{x} \pm t_{0.025} \frac{s}{\sqrt{n}} = 0.35 \pm (2.0049) \frac{0.0770}{\sqrt{55}}$

$= (0.3292, 0.3708)$

c. There is evidence to suggest the percentage of tannin is different in apples from fertilized and unfertilized trees. The CIs do not overlap.

8.137 a. $\bar{x} \pm t_{0.025} \frac{s}{\sqrt{n}} = 4.662 \pm (2.1098) \frac{0.298}{\sqrt{18}}$

$= (4.5138, 4.8102)$

b. $\bar{x} \pm t_{0.025} \frac{s}{\sqrt{n}} = 2.079 \pm (2.0555) \frac{0.309}{\sqrt{27}}$

$= (1.9568, 2.2012)$

c. There is no evidence that the population mean mercury concentration is greater than 5 for either group. Both CIs lie entirely below 5.

d. There is evidence to suggest the mean mercury concentration is different for the two groups. The CIs do not overlap.

8.139 a. $\hat{p} \pm z_{0.025} \sqrt{\frac{\hat{p}(1-\hat{p})}{n}}$

$= 0.3517 \pm (1.96) \sqrt{\frac{(0.3517)(0.6483)}{145}}$

$= (0.2740, 0.4294)$

b. $\hat{p} \pm z_{0.025} \sqrt{\frac{\hat{p}(1-\hat{p})}{n}}$

$= 0.3985 \pm (1.96) \sqrt{\frac{(0.3985)(0.6015)}{527}}$

$= (0.3567, 0.4403)$

c. There is no evidence to suggest the proportion of males with heart disease is different for those who donate blood and those who do not. The CIs overlap.

Chapter 9: Hypothesis Tests Based on a Single Sample

Section 9.1: The Parts of a Hypothesis Test and Choosing the Alternative Hypothesis

9.1 a. Valid, null hypothesis   **b.** Invalid, $\widehat{p}$ is a statistic.   **c.** Invalid, s is a statistic.   **d.** Invalid, $\overline{x}$ is a statistic.   **e.** Valid, alternative hypothesis **f.** Valid, alternative hypothesis   **g.** Invalid, $\widetilde{x}$ is a statistic.   **h.** Valid, null hypothesis

9.3 a. Valid   **b.** Invalid. The alternative hypothesis should be H_a: $\mu > 9.7$.   **c.** Invalid. The alternative hypothesis should involve 98.6, for example, H_a: $\sigma^2 \neq 98.6$.   **d.** Valid

9.5 a. Valid   **b.** Valid   **c.** Invalid. The null hypothesis should be stated so that μ (a parameter) equals a single value.   **d.** Valid.

9.7 Let $\mu =$ the population mean cumulative SAT score. H_0: $\mu = 1151$, H_a: $\mu > 1151$

9.9 Let $\mu =$ the population mean number of acres burned during wildfires.
H_0: $\mu = 17{,}060$, H_a: $\mu < 17{,}060$

9.11 (a) is appropriate. The software company is looking for evidence that the mean age is greater than 25.

9.13 H_0: $p = 0.75$, H_a: $p > 0.75$

9.15 (c) is appropriate. The bus company is looking for evidence that the true proportion of parents who favor seat-belt installation is greater than 0.50.

9.17 H_0: $p = 0.35$, H_a: $p < 0.35$

9.19 H_0: $\sigma = 7$, H_a: $\sigma < 7$

9.21 H_0: $p = 0.60$, H_a: $p > 0.60$

9.23 (c) H_0: $p = 0.80$, H_a: $p > 0.80$ is appropriate. The City Council is looking for evidence that the true proportion of residents who favor the plan is greater than 80%.

9.25 H_0: $\widetilde{\mu} = 25.50$, H_a: $\widetilde{\mu} < 125.50$

Section 9.2: Hypothesis Test Errors

9.27 a. Correct decision   **b.** Type II error
c. Type II error   **d.** Correct decision

9.29 a. $\beta(11) > \beta(15)$. As the alternative value of μ moves farther from the hypothesized value, the probability of a type II error decreases. There is a better chance of detecting the difference.   **b.** The probability of a type II error decreases.

9.31 Since α and β are inversely related, if we set the probability of a type I error (α) to a very small value, then probability of a type II error (β) would be very large.

9.33 a. A type I error would be deciding that the mean age of the files is greater than 10 years when it is actually no more than 10 years. A Type II error would be failing to realize the mean age of the files is more than 10 years when it actually is.

b. A type II error is more serious from the warden's perspective because in this case the files are really very old and need to be archived.

c. A type I error is more serious from a state senator's point of view because in this case the files are really not that old and the money does not need to be spent to archive them.

9.35 a. H_0: $p = 0.08$, H_a: $p < 0.08$

b. A type II error is more serious for college officials because, in this case, the new academic policy is really working, but there is no evidence of that.

c. A type I error is more serious for students because, in this situation, the new academic policy is not working, but fewer students are showing up late for exams. This probably means the new policy would remain in effect, but it really isn't necessary.

9.37 a. H_0: $p = 0.60$, H_a: $p > 0.60$

b. A type I error would be deciding that more than 60% of the town residents favor the variance, when the true percentage is really 60% or less. A type II error would be failing to realize that more than 60% of the town residents favor the variance.

c. A type II error is more serious for the developer because, in this case, the residents are actually in favor of the extended structure, but the developer will not be able to move forward with the project.

d. A type I error is more serious for the city council member because, in this case, they believe the residents are in favor of the extended structure, but they really aren't.

9.39 a. H_0: $\mu = 1367$, H_a: $\mu > 1367$

b. A type I error would be deciding the mean stamp duty for first-time home buyers is more than £1367 when it is, at most, £1367. A type II error would be failing to realize that the mean stamp duty has risen to more than £1367.

9.41 a. H_0: $\sigma^2 = 15$, H_a: $\sigma^2 < 15$

b. A type I error would be deciding the variance was less than 15 when the true population variance is really 15 or more. A type II error would be failing to recognize that the population variance was less than 15.

c. A type I error is more serious for the NSF because they would commit more money for TM research when there is really no evidence TM decreases brain activity. A type II error would be more serious for TM researchers because this would make it difficult for them to obtain more research support (money) in the face of the lack of evidence of decreased brain activity when TM really works!

9.43 a. H_0: $p = 0.15$, H_a: $p > 0.15$

b. A type I error: Decide $p > 0.15$ when the true proportion is really 0.15 (or less). Type II error: Decide $p = 0.15$ (or less) when the true proportion is really greater than 0.15.

c. The probability of a type I error becomes smaller.

Section 9.3: Hypothesis Tests Concerning a Population Mean When σ Is Known

9.45 a. $Z = \frac{\overline{X} - 170}{15/\sqrt{38}}$

b. (i) $Z \leq -2.3263$ (ii) $Z \leq -1.96$
(iii) $Z \leq -1.6449$ (iv) $Z \leq -1.2816$
(v) $Z \leq -3.0902$ (vi) $Z \leq -3.7190$

9.47 a. $Z = \frac{\overline{X} - (-11)}{4.5/\sqrt{21}}$

b. (i) $|Z| \geq 2.5758$ (ii) $|Z| \geq 1.2816$
(iii) $|Z| \geq 1.96$ (iv) $|Z| \geq 1.6449$
(v) $|Z| \geq 3.2905$ (vi) $|Z| \geq 3.7190$

9.49 a. $\alpha = 0.05$ **b.** $\alpha = 0.10$ **c.** $\alpha = 0.005$
d. $\alpha = 0.001$ **e.** $\alpha = 0.20$ **f.** $\alpha = 0.02$

9.51 a. H_0: $\mu = 212$; H_a: $\mu > 212$;

TS: $Z = \frac{\overline{X} - \mu_0}{\sigma/\sqrt{n}}$; RR: $Z \geq 2.3263$

b. We assume the underlying population is normal and the population standard deviation (σ) is known.

c. $z = \frac{213.5 - 212}{2.88/\sqrt{25}} = 2.6042 \geq 2.3263$. There is evidence to suggest that the population mean is greater than 212.

9.53 a. H_0: $\mu = 365.25$; H_a: $\mu \neq 365.25$;

TS: $Z = \frac{\overline{X} - \mu_0}{\sigma/\sqrt{n}}$; RR: $|Z| \geq 1.96$

b. We assume the sample size is large and the population standard deviation (σ) is known.

c. $z = \frac{360 - 365.25}{22.3/\sqrt{48}} = 1.6311$. There is no evidence to suggest the population mean is different from 365.25.

9.55 H_0: $\mu = 51{,}500$; H_a: $\mu < 51{,}500$;

TS: $Z = \frac{\overline{X} - \mu_0}{\sigma/\sqrt{n}}$; RR: $Z \leq -2.3263$

$z = \frac{49{,}762 - 51{,}500}{3750/\sqrt{38}} = -2.8570 \leq -2.3263$

There is evidence to suggest the mean income per year of corporate communications workers has decreased.

9.57 H_0: $\mu = 295$; H_a: $\mu > 295$;

TS: $Z = \frac{\overline{X} - \mu_0}{\sigma/\sqrt{n}}$; RR: $Z \geq 2.3263$

$z = \frac{306.3 - 295}{52/\sqrt{48}} = 1.5056$

There is no evidence to suggest the mean length of international calls has increased. Therefore, there is no evidence to suggest that the campaign was successful.

9.59 a. H_0: $\mu = 35$; H_a: $\mu > 35$;

TS: $Z = \frac{\overline{X} - \mu_0}{\sigma/\sqrt{n}}$; RR: $Z \geq 2.3263$

$z = \frac{36.22 - 35}{\sqrt{5.7}/\sqrt{41}} = 3.2720 \geq 2.3263$

There is evidence to suggest the LOA is greater than 35 feet.

b. The answer does not change if $\alpha = 0.10$. The critical value in this case is 1.2816, which is even smaller than 2.3263.

9.61 a. H_0: $\mu = 2200$; H_a: $\mu < 2200$;

TS: $Z = \frac{\overline{X} - \mu_0}{\sigma/\sqrt{n}}$; RR: $Z \leq -1.6449$

$z = \frac{2089 - 2200}{358/\sqrt{37}} = -1.8860 \leq -1.6449$

There is evidence to suggest the mean caloric intake is below the daily energy requirement of 2200 calories.

b. If we use $\alpha = 0.01$, the rejection region is RR: $Z \leq -2.3263$. In this case, there is no evidence to suggest the mean caloric intake is below the daily energy requirement.

9.63 a. H_0: $\mu = 3.9$; H_a: $\mu > 3.9$;

TS: $Z = \frac{\overline{X} - \mu_0}{\sigma/\sqrt{n}}$; RR: $Z \geq 2.3263$

$z = \frac{4.65 - 3.9}{2.3/\sqrt{10}} = 1.0312$

There is no evidence to suggest the mean daily temperature in 2008 was higher than the long-term mean air temperature.

b. $z = \frac{4.65-3.9}{1.3/\sqrt{10}} = 1.8244$

The conclusion does not change.

9.65 a. H_0: $\mu = 12$; H_a: $\mu < 12$;

TS: $Z = \frac{\overline{X}-\mu_0}{\sigma/\sqrt{n}}$; RR: $Z \leq -1.6449$

$z = \frac{11.85-12}{0.26/\sqrt{12}} = -1.9985 \leq -1.6449$

There is evidence to suggest the true mean impact velocity is less than 12 m/s.

b. If we use $\alpha = 0.01$, then the rejection region is RR: $Z \leq -2.3263$. In this case, there is no longer evidence to suggest that the true mean impact velocity is less than 12 m/s.

9.67 a. H_0: $\mu = 225$; H_a: $\mu < 225$;

TS: $Z = \frac{\overline{X}-\mu_0}{\sigma/\sqrt{n}}$; RR: $Z \leq -1.6449$

$z = \frac{224.2-225}{3.6/\sqrt{52}} = -1.6025$

There is not enough evidence to suggest that the mean amount of fuel in these cartridges is less than the advertised amount.

b. We assumed the sample size is large and that the population standard deviation (σ) is known.

9.69 H_0: $\mu = 42$; H_a: $\mu \neq 42$;

TS: $Z = \frac{\overline{X}-\mu_0}{\sigma/\sqrt{n}}$; RR: $|Z| \geq 1.96$

$z = \frac{43.22-42}{7.6/\sqrt{75}} = 1.3902$

There is no evidence to suggest that the true mean height of mail boxes in Des Moines is different from 42 inches.

9.71 a. $P\left(\frac{\overline{X}-50}{7.5/\sqrt{25}} \geq 2.3263\right) = 0.01$

$\Rightarrow$ $P(\overline{X} \geq 53.49) = 0.01$

$\beta(54) = P(\overline{X} < 53.49) = P\left(Z < \frac{53.49-54}{7.5/\sqrt{25}}\right)$

$= P(Z < -0.34) = 0.3669$

b. $\beta(55) = P\left(Z < \frac{53.49-55}{7.5/\sqrt{25}}\right)$

$= P(Z < -1.01) = 0.1562$

$\beta(56) = P\left(Z < \frac{53.49-56}{7.5/\sqrt{25}}\right)$

$= P(Z < -1.67) = 0.0475$

c. $P\left(\frac{\overline{X}-50}{7.5/\sqrt{25}} \geq 1.96\right) = 0.025$

$\Rightarrow$ $P(\overline{X} \geq 52.94) = 0.025$

$\beta(54) = P(\overline{X} < 52.94) = P\left(Z < \frac{52.94-54}{7.5/\sqrt{25}}\right)$

$= P(Z < -0.71) = 0.2398$

d. $\beta(55) = P\left(Z < \frac{52.94-55}{7.5/\sqrt{25}}\right)$

$= P(Z < -1.37) = 0.0853$

$\beta(56) = P\left(Z < \frac{52.94-56}{7.5/\sqrt{25}}\right)$

$= P(Z < -2.04) = 0.0207$

9.73 a. H_0: $\mu = 23.625$; H_a: $\mu \neq 23.625$;

TS: $Z = \frac{\overline{X}-\mu_0}{\sigma/\sqrt{n}}$; RR: $|Z| \geq 1.96$

$z = \frac{23.7-23.625}{0.15/\sqrt{10}} = 1.5811$

There is no evidence to suggest that the mean is less than 23.625, that the assembly line should be shut down.

b. $P\left(-1.96 \leq \frac{\overline{X}-23.625}{0.15/\sqrt{10}} \leq 1.96\right) = 0.025$

$\Rightarrow$ $P(23.532 \leq \overline{X} \leq 23.718) = 0.025$

9.75 a. H_0: $\mu = 3.32$; H_a: $\mu \neq 3.32$;

TS: $Z = \frac{\overline{X}-\mu_0}{\sigma/\sqrt{n}}$; RR: $|Z| \geq 2.5758$

$z = \frac{4.036-3.32}{2.8/\sqrt{28}} = 1.3531$

There is no evidence to suggest that the mean ice thickness has changed from 3.32 meters.

b. $P\left(\frac{\overline{X}-3.32}{2.8/\sqrt{28}} \geq 2.5758\right) = 0.005$

$\Rightarrow$ $P(\overline{X} \leq 4.6830) = 0.005$

$P\left(\frac{\overline{X}-3.32}{2.8/\sqrt{28}} \leq -2.5758\right) = 0.005$

$\Rightarrow$ $P(\overline{X} \leq 1.9570) = 0.005$

$\beta(3.0) = P(1.9570 \leq \overline{X} \leq 4.6830)$

$= P\left(\frac{1.9570-3.0}{2.8/\sqrt{28}} \leq Z \leq \frac{4.6830-3.0}{2.8/\sqrt{28}}\right)$

$= P(-1.97 \leq Z \leq 3.18)$

$= P(Z \leq 3.18) - P(Z \leq -1.97)$

$= 0.9993 - 0.0244 = 0.9749$

c. Illustration of the probability in part (b):

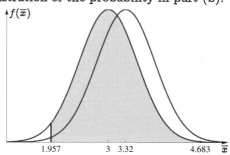

d. $\beta(2.8) = P(1.9570 \leq \overline{X} \leq 4.6830)$

$= P\left(\frac{1.9556-2.8}{2.8/\sqrt{28}} \leq Z \leq \frac{4.6844-2.8}{2.8/\sqrt{28}}\right)$

$= P(-1.60 \leq Z \leq 3.56)$

$= P(Z \leq 3.56) - P(Z \leq -1.59)$

$= 0.9998 - 0.0559 = 0.9439$

9.77 a. H_0: $\mu = 1250$; H_a: $\mu > 1250$;

TS: $Z = \frac{\overline{X} - \mu_0}{\sigma/\sqrt{n}}$; RR: $Z \geq 2.3263$

$z = \frac{1305 - 1250}{155/\sqrt{45}} = 2.3803 \geq 2.3263$

There is evidence to suggest that the population mean square footage is larger than 1250 during a sale.

b. $P\left(\frac{\overline{X} - 1250}{155/\sqrt{45}} \geq 2.3263\right) = 0.01$

$\Rightarrow P(\overline{X} \geq 1303.752) = 0.01$

$\beta(1330) = P(\overline{X} \leq 1303.752)$

$= P\left(Z \leq \frac{1303.752 - 1330}{155/\sqrt{45}}\right)$

$= P(Z < -1.14) = 0.1217$

c. Illustration:

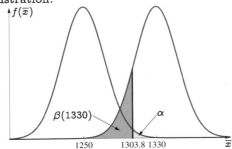

9.79 a. $P\left(\frac{\overline{X} - 1050}{3.7/\sqrt{18}} \geq 2.5758\right) = 0.005$

$\Rightarrow P(\overline{X} \geq 1052.2464) = 0.005$

$P\left(\frac{\overline{X} - 1050}{3.7/\sqrt{18}} \leq -2.5758\right) = 0.005$

$\Rightarrow P(\overline{X} \leq 1047.7536) = 0.005$

Critical values: 1047.7536 and 1052.2464

b. $H_0: \mu = 1050$; $H_a: \mu \neq 1050$;

TS: $Z = \frac{\overline{X} - \mu_0}{\sigma/\sqrt{n}}$; RR: $|Z| \geq 2.5758$

$z = \frac{1049 - 1050}{3.7/\sqrt{18}} = 1.1467$

There is no evidence to suggest that the population mean is different from 1050. The assembly line should not be shut down.

Alternatively, $\overline{x} = 1049$ is not in the rejection region in the $\overline{X}$ world. Therefore, this leads to the same conclusion.

Section 9.4: p Values

9.81 a. Do not reject   **b.** Reject   **c.** Do not reject   **d.** Do not reject   **e.** Reject   **f.** Do not reject

9.83 a. 0.0202   **b.** 0.0764   **c.** 0.0006   **d.** 0.2514   **e.** 0.00002327   **f.** 0.5987

9.85 a. $p = 0.0764$; do not reject H_0
b. $p = 0.0202$; reject H_0

c. $p = 0.0801$; reject H_0
d. $p = 0.0009$; reject H_0
e. $p = 0.1230$; do not reject H_0
f. $p = 0.0188$; do not reject H_0

9.87 a. $p = 0.2000$; do not reject H_0
b. $p = 0.1671$; do not reject H_0
c. $p = 0.0021$; reject H_0
d. $p = 0.0068$; do not reject H_0
e. $p = 0.7288$; do not reject H_0
f. $p = 0.0094$; reject H_0

9.89 $H_0: \mu = 87.6$; $H_a: \mu > 87.6$;

TS: $Z = \frac{\overline{X} - \mu_0}{\sigma/\sqrt{n}}$

$z = \frac{92.8 - 87.6}{15.7/\sqrt{43}} = 2.1719 \Rightarrow p = 0.0149$

There is some evidence to suggest that the mean quarterback rating is greater than 87.6.

9.91 a. $H_0: \mu = 30$; $H_a: \mu > 30$;

TS: $Z = \frac{\overline{X} - \mu_0}{\sigma/\sqrt{n}}$; RR: $Z \geq 2.3263$

$z = \frac{32.2 - 30}{5.7/\sqrt{58}} = 2.9394 \geq 2.3263$

There is evidence to suggest that the mean time to complete the safety checkup is more than 30 minutes.

b. $p = P(Z \geq 2.94) = 0.0016$

9.93 $H_0: \mu = 40$; $H_a: \mu \neq 40$;

TS: $Z = \frac{\overline{X} - \mu_0}{\sigma/\sqrt{n}}$

$z = \frac{40.913 - 40}{2/\sqrt{8}} = 1.2939 \Rightarrow p = 0.1969$

There is no evidence to suggest that the population mean weight of the blocks is different from 40 pounds.

9.95 $H_0: \mu = 185$; $H_a: \mu \neq 185$;

TS: $Z = \frac{\overline{X} - \mu_0}{\sigma/\sqrt{n}}$

$z = \frac{185.447 - 185}{2.7/\sqrt{32}} = 0.94 \Rightarrow p = 0.3472$

There is no evidence to suggest that the true mean weight of recycled shingles is different from 185 lbs/ft^3.

9.97 a. $H_0: \mu = 60$; $H_a: \mu < 60$;

TS: $Z = \frac{\overline{X} - \mu_0}{\sigma/\sqrt{n}}$;

$z = \frac{54.5998 - 60}{0.8/\sqrt{40}} = -4.27 \Rightarrow p = 0.0000098$

There is strong evidence to suggest that the mean drying time for highway paint is less than 60 seconds.

b. Yes, there is very strong evidence to suggest the paint dries in less than 60 seconds because the p value is so small.

c. p value illustration:

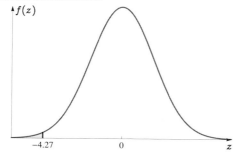

9.99 a. H_0: $\mu = 85$; H_a: $\mu < 85$;

TS: $Z = \frac{\overline{X} - \mu_0}{\sigma/\sqrt{n}}$;

$z = \frac{84.88 - 85}{5.6/\sqrt{60}} = -0.17 \Rightarrow p = 0.4341$

There is no evidence to suggest that the mean decibel level produced by smoke alarms in this city is less than 85.

b. The smallest significance level at which this hypothesis test would be significant is the p value, 0.4341.

Section 9.5: Hypothesis Tests Concerning a Population Mean When σ Is Unknown

9.101 a. TS: $T = \frac{\overline{X} - \mu_0}{S/\sqrt{n}}$

b. (i) $T \leq -2.6245$ (ii) $T \leq -4.5869$
(iii) $T \leq -1.7247$ (iv) $T \leq -1.3195$
(v) $T \leq -7.1732$ (vi) $T \leq -4.2340$

9.103 a. $\alpha = 0.025$ **b.** $\alpha = 0.001$ **c.** $\alpha = 0.01$
d. $\alpha = 0.001$

9.105 a. $\alpha = 0.1$ **b.** $\alpha = 0.01$ **c.** $\alpha = 0.002$
d. $\alpha = 0.0002$

9.107 a. $0.025 \leq p \leq 0.05$ **b.** $p > 0.20$
c. $0.01 \leq p \leq 0.025$ **d.** $0.0005 \leq p \leq 0.10$

9.109 a. H_0: $\mu = 1.618$; H_a: $\mu < 1.618$;

TS: $T = \frac{\overline{X} - \mu_0}{S/\sqrt{n}}$; RR: $T \leq -1.7291$

b. $t = \frac{1.5 - 1.618}{0.45/\sqrt{20}} = -1.1727$

There is no evidence to suggest that the mean is less than 1.618.

c. $p = P(T \leq -1.1727) = 0.1277$

9.111 a. H_0: $\mu = 9.96$; H_a: $\mu \neq 9.96$;

TS: $T = \frac{\overline{X} - \mu_0}{S/\sqrt{n}}$; RR: $|T| \geq 3.4210$

b. $t = \frac{9.04 - 9.96}{1.2/\sqrt{28}} = -4.0568 \leq -3.4210$

There is evidence to suggest that the mean is different from 9.96.

c. $p = 2P(T \leq -3.4210) = 0.0004$

9.113 H_0: $\mu = 871$; H_a: $\mu > 871$;

TS: $T = \frac{\overline{X} - \mu_0}{S/\sqrt{n}}$; RR: $T \geq 1.7959$

$t = \frac{885.7 - 871}{52/\sqrt{12}} = 0.9793$

There is no evidence to suggest that the mean weight of the moose taken from Island Pond is greater than 871 pounds.

9.115 H_0: $\mu = 31.9$; H_a: $\mu < 31.9$;

TS: $T = \frac{\overline{X} - \mu_0}{S/\sqrt{n}}$; RR: $T \leq -2.4851$

$t = \frac{30.088 - 31.9}{4.433/\sqrt{26}} = -2.0842$

There is no evidence to suggest the mean is less than 31.9.

9.117 H_0: $\mu = 40$; H_a: $\mu > 40$;

TS: $T = \frac{\overline{X} - \mu_0}{S/\sqrt{n}}$; RR: $T \geq 2.6503$

$t = \frac{44.3929 - 40}{3.384/\sqrt{14}} = 4.8568 \geq 2.6503$

There is strong evidence to suggest that the mean number of hours worked during the last week by white-collar employees was greater than 40.

9.119 H_0: $\mu = 350$; H_a: $\mu > 350$;

TS: $T = \frac{\overline{X} - \mu_0}{S/\sqrt{n}}$; RR: $T \geq 1.9432$

$t = \frac{353.8 - 350}{5.6/\sqrt{7}} = 1.7953$

There is no evidence to suggest that the mean power consumption for this model plasma TV fails to meet the design specifications.

9.121 a. H_0: $\mu = 25$; H_a: $\mu > 25$;

TS: $T = \frac{\overline{X} - \mu_0}{S/\sqrt{n}}$; RR: $T \geq 2.4377$

$t = \frac{28.0642 - 25}{1.5556/\sqrt{36}} = 2.8889 \geq 2.4377$

There is evidence to suggest that the population mean lead content in coral calcium is greater than the maximum allowance set by the CDC (25 mg).

b. $0.001 \leq p \leq 0.005$

9.123 H_0: $\mu = 55.5$; H_a: $\mu > 55.5$;

TS: $T = \frac{\overline{X} - \mu_0}{S/\sqrt{n}}$; RR: $T \geq 1.7959$

$t = \frac{70.3417 - 55.5}{52.9/\sqrt{12}} = 0.9727$

There is no evidence to suggest that the mean spending on Internet advertising has increased.

9.125 a. H_0: $\mu = 2$; H_a: $\mu \neq 2$;

TS: $T = \frac{\overline{X} - \mu_0}{S/\sqrt{n}}$; RR: $|T| \geq 2.8453$

$t = \frac{2.0619 - 2}{0.2622/\sqrt{21}} = 1.0819$

There is no evidence to suggest that the mean calcium concentration is different from 2 mg/L.

b. $0.2 \leq p \leq 0.4$

9.127 a. H_0: $\mu = 15$; H_a: $\mu < 15$;

TS: $T = \frac{\overline{X} - \mu_0}{S/\sqrt{n}}$; RR: $T \leq -2.2281$

$t = \frac{14.35 - 15}{3.75/\sqrt{11}} = -0.5749$

There is no evidence to suggest that the mean savings is less than \$15.

b. With such a small sample size, it is reasonable to obtain a sample mean of \$14.35 if the population men is really \$15 due to natural sampling variability.

c. $p > 0.20$

9.129 a. H_0: $\mu = 1659$; H_a: $\mu < 1659$;

TS: $T = \frac{\overline{X} - \mu_0}{S/\sqrt{n}}$; RR: $T \leq -1.7139$

$t = \frac{1652.83 - 1659}{17/\sqrt{24}} = -1.7733 \leq -1.7139$

There is evidence to suggest that the population mean monthly milk production per cow has decreased. The data was not obtained from a randomized, comparative experiment, so we cannot conclude that the heat wave caused this decrease in milk production.

b. $0.025 \leq p \leq 0.05$

9.131 a. H_0: $\mu = 0$; H_a: $\mu \neq 0$;

TS: $T = \frac{\overline{X} - \mu_0}{S/\sqrt{n}}$; RR: $|T| \geq 2.8609$

$t = \frac{-0.4205 - 0}{2.7246/\sqrt{20}} = -0.6902$.

There is no evidence to suggest the mean is different from 0. $p > 0.40$

b. There is no evidence to suggest prevailing drought or wet conditions.

9.133 a. H_0: $\mu = 1000$; H_a: $\mu > 1000$;

TS: $T = \frac{\overline{X} - \mu_0}{S/\sqrt{n}}$; RR: $T \geq 1.7033$

$t = \frac{1003.7 - 1000}{20.7/\sqrt{28}} = 0.9399$

There is no evidence to suggest that the population mean fluoride concentration per tube of toothpaste is greater than 1000 ppmF.

b. $0.1 \leq p \leq 0.2$

9.135 a. H_0: $\mu = 2748$; H_a: $\mu > 2748$;

TS: $T = \frac{\overline{X} - \mu_0}{S/\sqrt{n}}$; RR: $T \geq 2.7638$

$t = \frac{3022 - 2748}{457/\sqrt{11}} = 1.9885$

There is no evidence to suggest that the population mean credit card debt for undergraduates is greater than 2748.

b. $0.025 \leq p \leq 0.05$

c. Solve $T = \frac{3022 - 2748}{457/\sqrt{n}} \geq t_{0.01}$ for n.

Through trial and error, $n \geq 19$.

9.137 a. H_0: $\lambda = 11$; H_a: $\lambda > 11$;

TS: $Z = \frac{x - \lambda_0}{\sqrt{\lambda_0}}$; RR: $Z \geq 2.3263$

b. $Z = \frac{18 - 11}{\sqrt{11}} = 2.1106$

There is no evidence to suggest that the mean number of dog bites per day is greater than 11.

c. $p = P(Z \geq 2.11) = 0.0174$

Section 9.6: Large-Sample Hypothesis Tests Concerning a Population Proportion

9.139

RR	Value of TS	Conclusion
a. $Z \geq 1.6449$	0.3033	Do not reject H_0
b. $Z \geq 1.2816$	1.4882	Reject H_0
c. $Z \geq 2.3263$	2.3327	Reject H_0
d. $Z \geq 1.9600$	0.6872	Do not reject H_0
e. $Z \geq 2.3263$	0.3307	Do not reject H_0

9.141

RR	Value of TS	Conclusion
a. $|Z| \geq 2.2414$	−2.0142	Do not reject H_0.
b. $|Z| \geq 2.3263$	2.3451	Reject H_0
c. $|Z| \geq 1.9600$	−2.0294	Reject H_0
d. $|Z| \geq 2.8070$	−1.4025	Do not reject H_0
e. $|Z| \geq 2.5758$	2.6455	Reject H_0

9.143

Value of TS	p value	Conclusion
a. −2.0285	0.0213	Reject H_0
b. 0.0574	0.5229	Do not reject H_0
c. −0.7611	0.2233	Do not reject H_0
d. −2.2166	0.0133	Reject H_0
e. −2.6187	0.0044	Reject H_0

9.145 a. $n = 500$, $x = 16$, $p_0 = 0.02$

b. $500(0.02) = 10 \geq 5$; $500(0.98) = 490 \geq 5$
The large sample test is appropriate.

c. H_0: $p = 0.02$; H_a: $p > 0.02$;

TS: $Z = \frac{\hat{P} - p_0}{\sqrt{\frac{p_0(1 - p_0)}{n}}}$; RR: $Z \geq 1.6449$

$z = \frac{0.032 - 0.02}{\sqrt{\frac{(0.02)(0.98)}{500}}} = 1.9166 \geq 1.6449$

There is evidence to suggest that the proportion of children who eat the recommended number of servings each day has increased.

d. $p = P(Z \geq 1.9166) = 0.0276$

9.147 a. $n = 225$, $x = 189$, $p_0 = 0.90$

b. $225(0.90) = 202.5 \geq 5$; $225(0.10) = 22.5 \geq 5$
The large-sample test is appropriate.

c. H_0: $p = 0.90$; H_a: $p \neq 0.90$;

TS: $Z = \dfrac{\widehat{P} - p_0}{\sqrt{\frac{p_0(1-p_0)}{n}}}$; RR: $|Z| \geq 1.96$

$z = \dfrac{0.84 - 0.90}{\sqrt{\frac{(0.90)(0.10)}{225}}} = -3.0 \leq -1.96$

There is evidence to suggest that the proportion of homeschooled children who attend college is different from 0.90.

d. $p = 2P(Z \leq -3.0) = 0.0027$

9.149 H_0: $p = 0.30$; H_a: $p > 0.30$;

TS: $Z = \dfrac{\widehat{P} - p_0}{\sqrt{\frac{p_0(1-p_0)}{n}}}$; RR: $Z \geq 3.0902$

$z = \dfrac{0.379 - 0.30}{\sqrt{\frac{(0.30)(0.70)}{470}}} = 3.7243 \geq 3.0902$

There is evidence to suggest that the proportion of veterans using this benefit has increased.

9.151 H_0: $p = 0.79$; H_a: $p < 0.79$;

TS: $Z = \dfrac{\widehat{P} - p_0}{\sqrt{\frac{p_0(1-p_0)}{n}}}$; RR: $Z \leq -2.3263$

$z = \dfrac{0.7632 - 0.79}{\sqrt{\frac{(0.79)(0.21)}{321}}} = -1.1771$

There is no evidence to suggest that the proportion of professionals searching for a job closer to home has decreased.

9.153 H_0: $p = 0.40$; H_a: $p < 0.40$;

TS: $Z = \dfrac{\widehat{P} - p_0}{\sqrt{\frac{p_0(1-p_0)}{n}}}$; RR: $Z \leq -2.3263$

$z = \dfrac{0.3387 - 0.40}{\sqrt{\frac{(0.40)(0.60)}{375}}} = -2.4244 \leq -2.3263$

There is evidence to suggest that fewer than 40% of all residents are satisfied with the local government. This politician should enter the race for mayor.

9.155 H_0: $p = 0.10$; H_a: $p < 0.10$;

TS: $Z = \dfrac{\widehat{P} - p_0}{\sqrt{\frac{p_0(1-p_0)}{n}}}$; RR: $Z \leq -1.6449$

$z = \dfrac{0.12 - 0.10}{\sqrt{\frac{(0.10)(0.90)}{100}}} = 0.6667$

There is no evidence to suggest that the proportion of rent-controlled apartments is less than 0.10.

9.157 H_0: $p = 0.64$; H_a: $p \neq 0.64$;

TS: $Z = \dfrac{\widehat{P} - p_0}{\sqrt{\frac{p_0(1-p_0)}{n}}}$; RR: $|Z| \geq 1.96$

$z = \dfrac{0.7043 - 0.64}{\sqrt{\frac{(0.64)(0.36)}{328}}} = 2.4249 \geq 1.96$

There is evidence to suggest that the proportion of female principals in Florida public schools is different from 0.64.

9.159 a. H_0: $p = 0.95$; H_a: $p < 0.95$;

TS: $Z = \dfrac{\widehat{P} - p_0}{\sqrt{\frac{p_0(1-p_0)}{n}}}$; RR: $Z \leq -1.6449$

$z = \dfrac{0.915 - 0.95}{\sqrt{\frac{(0.95)(0.05)}{200}}} = -2.2711 \leq -1.6449$

There is evidence to suggest that the proportion of all tractor batteries that last at least three years is less than 0.95.

b. $p = P(Z \leq -2.2711) = 0.0116$

c. Since there is good evidence to suggest that less than 95% of all batteries last at least three years, the company should not implement the new warranty.

9.161 a. H_0: $p = 0.29$; H_a: $p < 0.29$;

TS: $Z = \dfrac{\widehat{P} - p_0}{\sqrt{\frac{p_0(1-p_0)}{n}}}$; RR: $Z \leq -2.3284$

$z = \dfrac{0.26 - 0.29}{\sqrt{\frac{(0.29)(0.71)}{1000}}} = -2.0907$

There is no evidence to suggest that the proportion of U.S. adults who have not filled a prescription because of cost is less than 0.29.

b. $P\left(\dfrac{\widehat{P} - 0.29}{\sqrt{\frac{(0.29)(0.71)}{1000}}} \leq -2.3263 \right) = 0.01$

$\Rightarrow P(\widehat{P} \leq 0.2566) = 0.01$

$\beta(0.26) = P(\widehat{P} \geq 0.2566) = P\left(Z \geq \dfrac{0.2566 - 0.26}{\sqrt{\frac{(0.26)(0.74)}{1000}}} \right)$

$= P(Z \geq -0.0005) = 0.5002$

c. Needed critical value is 1.2816:
$P(Z \geq 1.2816) = 0.10$

Solve: $\dfrac{0.2566 - p_\alpha}{\sqrt{\frac{p_\alpha(1-p_\alpha)}{1000}}} = 1.2816$

$\Rightarrow (0.2566 - p_\alpha)^2 = (1.2816)^2 \dfrac{(p_\alpha)(1-p_\alpha)}{1000}$

$\Rightarrow 0.2566^2 - 0.5132 p_\alpha + p_\alpha^2 = \dfrac{1.2816^2}{1000}(p_\alpha - p_\alpha^2)$

$\Rightarrow \left(1 + \dfrac{1.2816^2}{1000} \right) p_\alpha^2 - \left(0.5132 - \dfrac{1.2816^2}{1000} \right) p_\alpha$
$+ 0.2566^2 = 0$

$\Rightarrow p_\alpha = 0.2393$

d. Solve for n: $\dfrac{0.2566 - 0.24}{\sqrt{\frac{(0.24)(0.76)}{n}}} = 1.96$

$\Rightarrow n = 2542.85$

Since n must be an integer, $n \geq 2543$.

Section 9.7: Hypothesis Tests Concerning a Population Variance or Standard Deviation

9.163 a. $X^2 = \dfrac{(n-1)S^2}{\sigma_0^2}$

b. **(i)** $X^2 \geq 19.6751$ **(ii)** $X^2 \geq 31.5264$
(iii) $X^2 \geq 40.2894$ **(iv)** $X^2 \geq 37.9159$
(v) $X^2 \geq 20.5150$ **(vi)** $X^2 \geq 42.5793$

9.165 a. $X^2 = \dfrac{(n-1)S^2}{\sigma_0^2}$

b. (i) $X^2 \leq 23.6543$ or $X^2 \geq 58.1201$
(ii) $X^2 \leq 13.7867$ or $X^2 \geq 53.6720$
(iii) $X^2 \leq 5.8957$ or $X^2 \geq 49.0108$
(iv) $X^2 \leq 5.2293$ or $X^2 \geq 30.5779$
(v) $X^2 \leq 10.3909$ or $X^2 \geq 56.8923$
(vi) $X^2 \leq 1.0636$ or $X^2 \geq 7.7794$

9.167 a. $\alpha = 0.0005$ **b.** $\alpha = 0.01$ **c.** $\alpha = 0.025$
d. $\alpha = 0.005$

9.169 a. $0.01 \leq p \leq 0.025$ **b.** $0.05 \leq p \leq 0.10$
c. $0.001 \leq p \leq 0.005$ **d.** $p \leq 0.0001$

9.171 a. $0.02 \leq p \leq 0.05$ **b.** $0.0002 \leq p \leq 0.001$
c. $0.05 \leq p \leq 0.10$ **d.** $0.001 \leq p \leq 0.002$

9.173 a. H_0: $\sigma^2 = 36.8$; H_a: $\sigma^2 \neq 36.8$;
TS: $X^2 = \frac{(n-1)S^2}{\sigma_0^2}$; RR: $X^2 \leq 6.2621$ or $X^2 \geq 27.4884$
b. $\chi^2 = \frac{(15)(105.863)}{36.8} = 43.1508 \geq 27.4882$
There is evidence to suggest that the population variance is different from 36.8.
c. $0.0002 \leq p \leq 0.01$

9.175 H_0: $\sigma^2 = 0.25$; H_a: $\sigma^2 > 0.25$;
TS: $X^2 = \frac{(n-1)S^2}{\sigma_0^2}$; RR: $X^2 \geq 48.6024$
$\chi^2 = \frac{(34)(0.558)^2}{0.25} = 42.3455$
There is no evidence to suggest that the population variance is larger than 0.25.

9.177 H_0: $\sigma^2 = 324$; H_a: $\sigma^2 > 324$;
TS: $X^2 = \frac{(n-1)S^2}{\sigma_0^2}$; RR: $X^2 \geq 24.7250$
$\chi^2 = \frac{(11)(21.56)^2}{324} = 15.7814$
There is no evidence to suggest that the population variance in stress is greater than 324. Therefore, there is no evidence to refute the manufacturer's claim.

9.179 H_0: $\sigma^2 = 62.5$; H_a: $\sigma^2 > 62.5$;
TS: $X^2 = \frac{(n-1)S^2}{\sigma_0^2}$; RR: $X^2 \geq 21.6660$
$\chi^2 = \frac{(9)(70.1)}{62.5} = 10.0994$
There is no evidence to suggest that the population variance in yeast slurry is greater than 62.5.

9.181 H_0: $\sigma^2 = 40,000$; H_a: $\sigma^2 < 40,000$;
TS: $X^2 = \frac{(n-1)S^2}{\sigma_0^2}$; RR: $X^2 \leq 44.0379$
$\chi^2 = \frac{(61)(191.184)^2}{40,000} = 55.7408$
There is no evidence to suggest that the population variance in lot size is less than 40,000. There is no

evidence to suggest that the population standard deviation in lot size is less than 200.

9.183 H_0: $\sigma^2 = 22.5$; H_a: $\sigma^2 < 22.5$;
TS: $X^2 = \frac{(n-1)S^2}{\sigma_0^2}$; RR: $X^2 \leq 23.2686$
$\chi^2 = \frac{(36)(15.6)}{22.5} = 24.96 \geq 23.2686$
There is evidence to suggest that the population variance in ride time is less than 22.5. Therefore, there is evidence to suggest that the bull riding has become less exciting.

9.185 a. H_0: $\sigma^2 = 0.36$; H_a: $\sigma^2 > 0.36$;
TS: $X^2 = \frac{(n-1)S^2}{\sigma_0^2}$; RR: $X^2 \geq 30.1435$
$\chi^2 = \frac{(19)(0.42)}{0.36} = 22.1667$
There is no evidence to suggest that the population variance in malic acid slurry is greater than 0.36.
b. $p > 0.10$

9.187 a. H_0: $\sigma^2 = 230$; H_a: $\sigma^2 > 230$;
TS: $X^2 = \frac{(n-1)S^2}{\sigma_0^2}$; RR: $X^2 \geq 38.8851$
$\chi^2 = \frac{(26)(13.93)^2}{230} = 21.9362$
There is no evidence to suggest that the population variance is greater than 230.
b. $p > 0.10$

9.189 a. H_0: $\sigma^2 = 7.5625$; H_a: $\sigma^2 > 7.5625$;
TS: $X^2 = \frac{(n-1)S^2}{\sigma_0^2}$; RR: $X^2 \geq 32.6706$
$\chi^2 = \frac{(21)(3.811)^2}{7.5625} = 40.3239$
There is evidence to suggest that the population variance in wingspan is greater than 7.5625.
b. $0.005 \leq p \leq 0.01$

9.191 a. H_0: $\sigma^2 = 1.56$; H_a: $\sigma^2 \neq 1.56$;
TS: $Z = \frac{S^2 - \sigma_0^2}{\sqrt{2}\sigma_0^2/\sqrt{n-1}}$; RR: $|Z| \geq 1.96$
b. $z = \frac{2.2 - 1.56}{\sqrt{2}(1.56)/\sqrt{47}} = 1.9888 \geq 1.96$
There is evidence to suggest the population variance in exchange rate is different from 1.56.
c. $p = 2P(Z \geq 1.9888) = 0.0467$
d. H_0: $\sigma^2 = 1.56$; H_a: $\sigma^2 \neq 1.56$;
TS: $X^2 = \frac{(n-1)S^2}{\sigma_0^2}$;
RR: $X^2 \leq 29.9562$ or $X^2 \geq 67.8206$
$\chi^2 = \frac{(47)(2.2)}{1.56} = 66.2821$
There is no evidence to suggest the population variance in exchange rate is different from 1.56.
$p = 2P(X^2 \geq 66.2821) = 0.0666$

The conclusion and p value are different.

Chapter Exercises

9.193 H_0: $\mu = 4$; H_a: $\mu \neq 4$;
TS: $Z = (\bar{x} - \mu_0)/(\sigma/\sqrt{n})$; RR: $|Z| \geq 2.5758$

a. $z = \frac{4.014-4}{0.05/\sqrt{40}} = 1.7709$

There is no evidence to suggest the population mean thickness is different from 4. The process should not be stopped.

b. $z = \frac{3.979-4}{0.05/\sqrt{40}} = -2.6563 \leq -2.5758$

There is evidence to suggest the population mean thickness is different from 4. The process should be stopped.

9.195 a. H_0: $\mu = 23$; H_a: $\mu > 23$;
TS: $Z = (\bar{x} - \mu_0)/(\sigma/\sqrt{n})$; RR: $Z \geq 2.3263$

$z = \frac{24.6-23}{2.28/\sqrt{18}} = 2.9773 \geq 2.3263$

There is evidence to suggest the population mean width is greater than 23.

b. $p = P(Z \geq 2.9773) = 0.0015$

9.197 H_0: $\mu = 3$; H_a: $\mu > 3$;
TS: $T = \frac{\bar{X} - \mu_0}{S/\sqrt{n}}$; RR: $T \geq 1.7613$

$t = \frac{3.63-3.0}{1.673/\sqrt{15}} = 1.4660$

There is no evidence to suggest the population mean FEF is greater than 3.

9.199 a. H_0: $p = 0.75$; H_a: $p < 0.75$;
TS: $Z = \frac{\hat{P} - p_0}{\sqrt{\frac{p_0(1-p_0)}{n}}}$; RR: $Z \leq -2.3263$

$z = \frac{0.70-0.75}{\sqrt{\frac{(0.75)(0.25)}{560}}} = -2.7325 \leq -2.3263$

There is evidence to suggest the population proportion of residents who favor additional power to tap phones is less than 0.75.

b. $p = P(Z \leq -2.7325) = 0.0031$

9.201 H_0: $p = 0.80$; H_a: $p < 0.80$;
TS: $Z = \frac{\hat{P} - p_0}{\sqrt{\frac{p_0(1-p_0)}{n}}}$; RR: $Z \leq -3.0902$

$z = \frac{0.7725-0.80}{\sqrt{\frac{(0.80)(0.20)}{3050}}} = -3.8025 \leq -3.0902$

There is evidence to suggest the population proportion of counterfeit goods from China has decreased.

9.203 H_0: $\sigma^2 = 0.0015$; H_a: $\sigma^2 > 0.0015$;
TS: $X^2 = (n-1)S^2/\sigma_0^2$; RR: $X^2 \geq 23.6848$

$\chi^2 = \frac{(14)(0.0026)}{0.0015} = 24.2667 \geq 23.6848$

There is evidence to suggest the population variance in diameter of viruses has increased.

9.205 H_0: $\sigma^2 = 62500^2$; H_a: $\sigma^2 > 62500^2$;
TS: $X^2 = (n-1)S^2/\sigma_0^2$; RR: $X^2 \geq 67.9852$

$\chi^2 = \frac{(36)(65268)^2}{62500^2} = 39.2593$

There is no evidence to suggest the population variance in blood platelet count has increased.

9.207 a. H_0: $\mu = 1800$; H_a: $\mu > 1800$;
TS: $Z = (\bar{x} - \mu_0)/(\sigma/\sqrt{n})$; RR: $Z \geq 1.6449$

$z = \frac{1852-1800}{202/\sqrt{36}} = 1.5446$

There is no evidence to suggest the population mean amount of ore extracted each day is greater than 1800. There is no evidence to suggest the new machinery has improved production.

b. $P(Z \geq 1.6449) = 0.05$

$\Rightarrow P\left(\frac{\bar{X}-1800}{202/6} \geq 1.6449\right) = 0.05$

$\Rightarrow P(\bar{X} \geq 1855.3783) = 0.05$

$\beta(1875) = P(\bar{X} \leq 1855.3783)$

$= P(Z \leq -0.5828) = 0.2800$

$\beta(1925) = P(\bar{X} \leq 1855.3783)$

$= P(Z \leq -2.068) = 0.0193$

9.209 a. H_0: $\mu = 40$; H_a: $\mu < 40$;
TS: $T = \frac{\bar{X} - \mu_0}{S/\sqrt{n}}$; RR: $T \leq -2.5083$

$t = \frac{38.63-40}{5.6/\sqrt{23}} = -1.1733$

There is no evidence to suggest the population mean brightness is less than 40.

b. $-1.3212 \leq -1.1733 \leq -0.8583$
$t_{0.10} \leq -1.1733 \leq t_{0.20}$
$0.10 \leq \quad p \quad \leq 0.20$

9.211 a. H_0: $p = 0.60$; H_a: $p < 0.60$;
TS: $Z = \frac{\hat{P} - p_0}{\sqrt{\frac{p_0(1-p_0)}{n}}}$; RR: $Z \leq -2.3263$

$z = \frac{0.475-0.60}{\sqrt{\frac{(0.60)(0.40)}{120}}} = -2.7951 \leq -2.3263$

There is evidence to suggest the population proportion of cast-iron pans with harmful bacteria is less than 0.60.

b. $p = P(Z \leq -2.7951) = 0.0026$

c. This is probably not a random sample. The people who brought their pans for testing self-selected.

9.213 a. H_0: $\sigma^2 = 3.1^2$; H_a: $\sigma^2 < 3.1^2$;
TS: $X^2 = (n-1)S^2/\sigma_0^2$; RR: $X^2 \leq 6.5706$

$\chi^2 = \frac{(14)(2.534)^2}{3.1^2} = 9.3511$

There is no evidence to suggest the population variance in b value has decreased.

b. $9.3511 \geq 7.7895 \Rightarrow 9.3511 \geq \chi^2_{0.90}$

$\Rightarrow p > 0.10$

Chapter 10: Confidence Intervals and Hypothesis Tests Based on Two Samples or Treatments

Section 10.1: Comparing Two Population Means, Independent Samples, Population Variances Known

10.1 a. $\mu_1 - \mu_2 = 0$ **b.** $\mu_1 - \mu_2 < 0$
c. $\mu_1 - \mu_1 \neq 7$ **d.** $\mu_1 - \mu_2 > -4$ **e.** $\mu_1 - \mu_2 \neq 0$
f. $\mu_1 - \mu_2 = 10$

10.3 a. H_0: $\mu_1 - \mu_2 = 0$; H_a: $\mu_1 - \mu_2 > 0$;
TS: $Z = \dfrac{(\overline{X}_1 - \overline{X}_2) - 0}{\sqrt{\frac{\sigma_1^2}{n_1} + \frac{\sigma_2^2}{n_2}}}$; RR: $Z \geq 1.6449$

b. $z = \dfrac{(17.5 - 16.2) - 0}{\sqrt{\frac{1.5^2}{18} + \frac{2.6^2}{26}}} = 2.0951 \geq 1.6449$

There is evidence to suggest population mean 1 is greater than population mean 2.

c. $p = P(Z \geq 2.0951) = 0.0181$

10.5 a. H_0: $\mu_1 - \mu_2 = 0$; H_a: $\mu_1 - \mu_2 \neq 0$;
TS: $Z = \dfrac{(\overline{X}_1 - \overline{X}_2) - 0}{\sqrt{\frac{\sigma_1^2}{n_1} + \frac{\sigma_2^2}{n_2}}}$; RR: $|Z| \geq 3.2905$

b. $z = \dfrac{(1025.6 - 1031.3) - 0}{\sqrt{\frac{225.3}{37} + \frac{107.6}{42}}} = -1.9379$

There is no evidence to suggest population mean 1 is different from population mean 2.

c. The normality assumption is not necessary since both sample sizes are large.

10.7 a. $\overline{X}_1 - \overline{X}_2$ is normal with mean
$25 - 10 = 15$, variance $= \frac{25}{15} + \frac{16}{21} = 2.4286$,
and standard deviation $= \sqrt{2.4286} = 1.5584$

b. Probability density function:

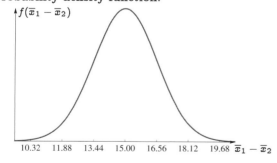

c. $P(\overline{X}_1 - \overline{X}_2 \geq 17) = P(Z \geq 1.28)$
$= 1 - P(Z \leq 1.28)$
$= 1 - 0.8997 = 0.1003$

d. $P(13.5 < \overline{X}_1 - \overline{X}_2 < 14.5)$
$= P(-0.96 < Z < -0.32)$
$= P(Z \leq -0.32) - P(Z \leq -0.96)$
$= 0.3745 - 0.1685 = 0.2060$

e. $P(\overline{X}_1 - \overline{X}_2 < 14)$
$= P(Z < -0.64) = 0.2611$

10.9 H_0: $\mu_1 - \mu_2 = 0$; H_a: $\mu_1 - \mu_2 \neq 0$;
TS: $Z = \dfrac{(\overline{X}_1 - \overline{X}_2) - 0}{\sqrt{\frac{\sigma_1^2}{n_1} + \frac{\sigma_2^2}{n_2}}}$; RR: $|Z| \geq 2.5758$

$z = \dfrac{(24.07 - 26.61) - 0}{\sqrt{\frac{16.81}{41} + \frac{10.24}{38}}} = -3.0814 \leq -2.5758$

There is evidence to suggest the mean Nordstrom gift-certificate value is different from the mean Macy's gift-certificate value.

10.11 a. $(\overline{x}_1 - \overline{x}_2) \pm z_{0.025} \sqrt{\frac{\sigma_1^2}{n_1} + \frac{\sigma_2^2}{n_2}}$

$= (99.79 - 100.36) \pm (1.96)\sqrt{\frac{1.5625}{16} + \frac{3.9024}{21}}$

$= (-1.6135, 0.4735)$

b. There is no evidence to suggest the mean power-output ratings for the two brands differ since 0 is in the CI.

10.13 a. H_0: $\mu_1 - \mu_2 = 5$; H_a: $\mu_1 - \mu_2 < 5$;
TS: $Z = \dfrac{(\overline{X}_1 - \overline{X}_2) - 5}{\sqrt{\frac{\sigma_1^2}{n_1} + \frac{\sigma_2^2}{n_2}}}$; RR: $Z \leq -2.3263$

$z = \dfrac{(17.99 - 13.26) - 5}{\sqrt{\frac{2.89}{42} + \frac{2.25}{38}}} = -0.7546$

There is no evidence to refute the claim; there is no evidence to suggest the difference in mean weights is less than 5 pounds.

b. $P(Z \leq -0.7546) = 0.2252$

c. The normality assumption is not necessary since both sample sizes are large.

10.15 a. $(\overline{x}_1 - \overline{x}_2) \pm z_{0.005} \sqrt{\frac{\sigma_1^2}{n_1} + \frac{\sigma_2^2}{n_2}}$

$= (102.9 - 97.6) \pm (2.5758)\sqrt{\frac{4.41}{34} + \frac{14.44}{35}}$

$= (3.4032, 7.1968)$

b. There is evidence to suggest that the mean Jensen sensitivity is greater than the mean Sennheiser sensitivity. 0 is not in the CI, and the CI is entirely above 0.

10.17 a. H_0: $\mu_1 - \mu_2 = 0$; H_a: $\mu_1 - \mu_2 > 0$;
TS: $Z = \dfrac{(\overline{X}_1 - \overline{X}_2) - 0}{\sqrt{\frac{\sigma_1^2}{n_1} + \frac{\sigma_2^2}{n_2}}}$; RR: $Z \geq 2.3263$

$z = \dfrac{(73.41 - 70.49) - 0}{\sqrt{\frac{13.69}{40} + \frac{21.16}{40}}} = 3.1283 \geq 2.3263$

There is evidence to suggest the population mean standby time for the Motorola phone is greater than the population mean standby time for the Uniden phone.

b. $P(Z \geq 3.1283) = 0.0009$

10.19 a. H_0: $\mu_1 - \mu_2 = 0$; H_a: $\mu_1 - \mu_2 \neq 0$;

TS: $Z = \dfrac{(\overline{X}_1 - \overline{X}_2) - 0}{\sqrt{\frac{\sigma_1^2}{n_1} + \frac{\sigma_2^2}{n_2}}}$; RR: $|Z| \geq 2.5758$

$z = \dfrac{(39.58 - 40.12) - 0}{\sqrt{\frac{2.47}{18} + \frac{0.87}{18}}} = -1.2536$

There is no evidence to suggest the population mean magnesium in each serving of baked beans and potatoes is different.

b. $z = \dfrac{(39.58 - 40.12) - 0}{\sqrt{\frac{2.47}{38} + \frac{0.87}{38}}} = -1.8214$

There is still no evidence to refute the claim.

$p = 2P(Z \leq -1.8214) = 0.0685$

c. Solve for n: $\dfrac{(39.58 - 40.12) - 0}{\sqrt{\frac{2.47}{n} + \frac{0.87}{n}}} = -2.5758$

$\Rightarrow n = 75.99 \Rightarrow n_1 = n_2 = 76$

10.21 a. H_0: $\mu_1 - \mu_2 = 0$; H_a: $\mu_1 - \mu_2 < 0$;

TS: $Z = \dfrac{(\overline{X}_1 - \overline{X}_2) - 0}{\sqrt{\frac{\sigma_1^2}{n_1} + \frac{\sigma_2^2}{n_2}}}$; RR: $Z \leq -2.3263$

$z = \dfrac{(29.21 - 33.90) - 0}{\sqrt{\frac{392.04}{60} + \frac{243.36}{75}}} = -1.4998$

There is no evidence to suggest the mean hourly wage for a plumber in Utica is less than the mean hourly wage for a plumber in Atlanta.

b. $p = P(Z \leq -1.4998) = 0.0668$

c. Both population variances are large.

Section 10.2: Comparing Two Population Means, Independent Samples, Normal Populations

10.23 a. H_0: $\mu_1 - \mu_2 = 0$; H_a: $\mu_1 - \mu_2 < 0$;

TS: $T = \dfrac{(\overline{X}_1 - \overline{X}_2) - 0}{\sqrt{S_p^2 \left(\frac{1}{n_1} + \frac{1}{n_2}\right)}}$; RR: $T \leq -1.7011$

$t = \dfrac{(49.6 - 50.2) - 0}{\sqrt{(192.8457)\left(\frac{1}{14} + \frac{1}{16}\right)}} = -0.1181$

There is no evidence to suggest μ_1 is less than μ_2.

b. $|t| = 0.1181$

$0.1181 \leq 0.8546 \Rightarrow 0.1181 \leq t_{0.20} \Rightarrow p > 0.20$

10.25 a. H_0: $\mu_1 - \mu_2 = 0$; H_a: $\mu_1 - \mu_2 \neq 0$;

TS: $T = \dfrac{(\overline{X}_1 - \overline{X}_2) - 0}{\sqrt{S_p^2 \left(\frac{1}{n_1} + \frac{1}{n_2}\right)}}$; RR: $|T| \geq 2.0484$

$t = \dfrac{(283.33 - 248.7) - 0}{\sqrt{(993.1813)\left(\frac{1}{15} + \frac{1}{15}\right)}} = 3.0096 \geq 2.0484$

There is evidence to suggest the two population means are different.

b. $|t| = 3.0096$

$2.7633 \leq 3.0096 \leq 3.4082$

$t_{0.005} \leq 3.0096 \leq t_{0.001}$

$0.001 \leq p/2 \leq 0.005$

$0.002 \leq p \leq 0.01$

10.27 a. df = 24   **b.** df = 15   **c.** df = 33

d. df = 62

10.29 a. H_0: $\mu_1 - \mu_2 = 0$; H_a: $\mu_1 - \mu_2 \neq 0$;

TS: $T = \dfrac{(\overline{X}_1 - \overline{X}_2) - 0}{\sqrt{S_p^2 \left(\frac{1}{n_1} + \frac{1}{n_2}\right)}}$; RR: $|T| \geq 2.0796$

$t = \dfrac{(76.83 - 66.8) - 0}{\sqrt{(143.1114)\left(\frac{1}{7} + \frac{1}{16}\right)}} = 1.8502$

There is no evidence to suggest that the two population means are different.

b. H_0: $\mu_1 - \mu_2 = 0$; H_a: $\mu_1 - \mu_2 \neq 0$;

TS: $T' = \dfrac{(\overline{X}_1 - \overline{X}_2) - 0}{\sqrt{\frac{S_1^2}{n_1} + \frac{S_2^2}{n_2}}}$; RR: $|T'| \geq 2.1009$, df = 18

$t' = \dfrac{(76.83 - 66.80) - 0}{\sqrt{\frac{3.30^2}{7} + \frac{14^2}{16}}} = 2.6994 \geq 2.1009$

There is evidence to suggest that the two population means are different.

c. The test in part (b) is more appropriate, where we assume the population variances are unequal. The sample standard deviations are not similar, suggesting that the population variances are unequal.

10.31 a. H_0: $\mu_1 - \mu_2 = 0$; H_a: $\mu_1 - \mu_2 \neq 0$;

TS: $T = \dfrac{(\overline{X}_1 - \overline{X}_2) - 0}{\sqrt{S_p^2 \left(\frac{1}{n_1} + \frac{1}{n_2}\right)}}$; RR: $|T| \geq 2.0687$

$t = \dfrac{(26.74 - 29.53) - 0}{\sqrt{(56.5459)\left(\frac{1}{11} + \frac{1}{14}\right)}} = -0.9209$

There is no evidence to suggest a difference in the population mean deviations from perfect flatness for these two rotor brands.

b. $|t| = 0.9209$, df = 23

$0.8575 \leq 0.9209 \leq 1.3195$

$t_{0.20} \leq 0.9209 \leq t_{0.10}$

$0.10 \leq p/2 \leq 0.20$

$0.20 \leq p \leq 0.40$

10.33 H_0: $\mu_1 - \mu_2 = 0$; H_a: $\mu_1 - \mu_2 < 0$;

TS: $T = \dfrac{(\overline{X}_1 - \overline{X}_2) - 0}{\sqrt{S_p^2 \left(\frac{1}{n_1} + \frac{1}{n_2}\right)}}$; RR: $T \leq -1.7011$

$t = \dfrac{(4.57 - 5.54) - 0}{\sqrt{(1.0967)^2 \left(\frac{1}{14} + \frac{1}{16}\right)}} = -2.4051 \leq -1.7011$

There is evidence to suggest the population mean size for rap MP3 files is less than the population mean size for jazz MP3 files.

10.35 a. H_0: $\mu_1 - \mu_2 = 0$; H_a: $\mu_1 - \mu_2 \neq 0$;

TS: $T = \dfrac{(\overline{X}_1 - \overline{X}_2) - 0}{\sqrt{S_p^2 \left(\frac{1}{n_1} + \frac{1}{n_2}\right)}}$; RR: $|T| \geq 2.6259$

$t = \dfrac{(2.5 - 2.8) - 0}{\sqrt{(0.49)^2 \left(\frac{1}{54} + \frac{1}{48}\right)}} = -3.0863 \leq -2.6259$

There is evidence to suggest the mean sagittal

diameter of women's biceps tendons is different from that of men's biceps tendons.

b. $(\overline{x}_1 - \overline{x}_2) \pm t_{0.025} \sqrt{s_p^2 \left(\frac{1}{n_1} + \frac{1}{n_2} \right)}$

$= (2.5 - 2.8) \pm (1.9840) \sqrt{(0.49)^2 \left(\frac{1}{54} + \frac{1}{48} \right)}$

$= (-0.4928, -0.1072)$

10.37 H_0: $\mu_1 - \mu_2 = 0$; H_a: $\mu_1 - \mu_2 > 0$;

TS: $T = \frac{(\overline{X}_1 - \overline{X}_2) - 0}{\sqrt{S_p^2 \left(\frac{1}{n_1} + \frac{1}{n_2} \right)}}$; RR: $T \geq 3.2614$

$t = \frac{(784 - 652) - 0}{\sqrt{(34.064)^2 \left(\frac{1}{27} + \frac{1}{25} \right)}} = 13.9613 \geq 3.2614$

There is evidence to suggest the mean amount men intend to spend is greater than the mean amount women intend to spend.

10.39 a. H_0: $\mu_1 - \mu_2 = 0$; H_a: $\mu_1 - \mu_2 > 0$;

TS: $T' = \frac{(\overline{X}_1 - \overline{X}_2) - 0}{\sqrt{\left(\frac{s_1^2}{n_1} + \frac{s_2^2}{n_2} \right)}}$; RR: $T' \geq 2.4922$

$t' = \frac{(331.4 - 298.7) - 0}{\sqrt{\frac{201.64}{10} + \frac{1190.25}{18}}} = 3.5202 \geq 2.4922$

There is evidence to suggest the population mean amount of coating for Fero is greater than the population mean amount of coating for Cintex.

b. $(\overline{x}_1 - \overline{x}_2) \pm t_{0.025} \sqrt{\frac{s_1^2}{n_1} + \frac{s_2^2}{n_2}}$

$= (331.4 - 298.7) \pm (2.0639) \sqrt{\frac{201.64}{10} + \frac{1190.25}{18}}$

$= (13.5281, 51.8719)$

10.41 H_0: $\mu_1 - \mu_2 = 0$; H_a: $\mu_1 - \mu_2 > 0$;

TS: $T' = \frac{(\overline{X}_1 - \overline{X}_2) - 0}{\sqrt{\left(\frac{s_1^2}{n_1} + \frac{s_2^2}{n_2} \right)}}$; RR: $T' \geq 2.4851$, df = 25

$t' = \frac{(0.02455 - 0.019867) - 0}{\sqrt{\frac{0.006428^2}{20} + \frac{0.002416^2}{15}}} = 2.9891 \geq 2.4851$

There is evidence to suggest the mean thickness of Aries wallpaper is greater than the mean thickness of all other wallpapers.

10.43 H_0: $\mu_1 - \mu_2 = 0$; H_a: $\mu_1 - \mu_2 \neq 0$;

TS: $T = \frac{(\overline{X}_1 - \overline{X}_2) - 0}{\sqrt{S_p^2 \left(\frac{1}{n_1} + \frac{1}{n_2} \right)}}$; RR: $|T| \geq 2.7500$

$t = \frac{(88.75 - 87.19) - 0}{\sqrt{(12.746)^2 \left(\frac{1}{16} + \frac{1}{16} \right)}} = 0.3467$

There is no evidence to suggest the population mean amount spent on gift cards per consumer is different on the East Coast and the West Coast.

10.45 H_0: $\mu_1 - \mu_2 = 0$; H_a: $\mu_1 - \mu_2 \neq 0$;

TS: $T = \frac{(\overline{X}_1 - \overline{X}_2) - 0}{\sqrt{S_p^2 \left(\frac{1}{n_1} + \frac{1}{n_2} \right)}}$; RR: $|T| \geq 2.0244$

$t = \frac{(15.65 - 14.7) - 0}{\sqrt{(2.9502)^2 \left(\frac{1}{20} + \frac{1}{20} \right)}} = 1.0183$

There is no evidence to suggest the population mean number of days in the hospital is different for Type A toxin and Type B toxin.

10.47 a. H_0: $\mu_1 - \mu_2 = 0$; H_a: $\mu_1 - \mu_2 < 0$;

TS: $T = \frac{(\overline{X}_1 - \overline{X}_2) - 0}{\sqrt{S_p^2 \left(\frac{1}{n_1} + \frac{1}{n_2} \right)}}$; RR: $T \leq -2.4121$

$t = \frac{(66.5990 - 66.6924) - 0}{\sqrt{(0.0968)^2 \left(\frac{1}{23} + \frac{1}{24} \right)}} = -3.3073 \leq -2.4121$

There is evidence to suggest the mean width of a $20 bill is greater than the mean width of a $1 bill.

b. H_0: $\mu_1 - \mu_2 = 0$; H_a: $\mu_1 - \mu_2 < 0$;

TS: $T' = \frac{(\overline{X}_1 - \overline{X}_2) - 0}{\sqrt{\left(\frac{s_1^2}{n_1} + \frac{s_2^2}{n_2} \right)}}$; RR: $T' \leq -2.4314$, df = 37

$t' = \frac{(66.5990 - 66.6924) - 0}{\sqrt{\frac{0.0132}{23} + \frac{0.0057}{24}}} = -3.28 \leq -2.4314$

There is evidence to suggest the mean width of a $20 bill is greater than the mean width of a $1 bill.

c. The sample means are relatively close, the sample variances are relatively close, and the sample sizes are relatively large.

10.49 a. H_0: $\mu_1 - \mu_2 = 0$; H_a: $\mu_1 - \mu_2 \neq 0$;

TS: $T' = \frac{(\overline{X}_1 - \overline{X}_2) - 0}{\sqrt{\left(\frac{s_1^2}{n_1} + \frac{s_2^2}{n_2} \right)}}$; RR: $|T'| \geq 1.9996$, df = 61

$t' = \frac{(147.6 - 147.8) - 0}{\sqrt{\frac{7.09}{30} + \frac{13.70}{35}}} = -0.2524$

There is no evidence to suggest the mean pressure required to open each valve is different.

b. $(\overline{x}_1 - \overline{x}_2) \pm t_{0.025} \sqrt{\frac{s_1^2}{n_1} + \frac{s_2^2}{n_2}}$

$= (147.6 - 147.8) \pm (1.9996) \sqrt{\frac{7.09}{30} + \frac{13.70}{35}}$

$= (-1.784, 1.3842)$

This CI is consistent with part (a); 0 is in the CI.

10.51 You should reject the null hypothesis roughly 5 out of every 100 trials. This is true no matter what value of σ_2 is used. Although, as σ_2 gets farther from σ_1, you should reject the null hypothesis more often. Robust means our results are still 95% accurate, even when the underlying assumptions are not met.

Section 10.3: Paired Data

10.53 a. The data are paired; the common characteristic is the affected arm of the patient with rotator cuff injuries. **b.** The data are paired; the common characteristic is the lathes. **c.** The data were obtained independently; the two populations are 20-year-old males and 70-year-old males.

d. The data are paired; the common characteristic is the files on a PDA.   **e.** The data were obtained independently; the two populations are frequent flyers on United Airlines and frequent flyers on Delta Airlines.

10.55 a. H_0: $\mu_D = 0$; H_a: $\mu_D < 0$;

TS: $T = \frac{\overline{D} - \Delta_0}{S_D/\sqrt{n}}$; RR: $T \leq -3.7469$

$t = \frac{-13.9 - 0}{29.4/\sqrt{5}} = -1.0551$

There is no evidence to suggest the population mean before treatment is less than the population mean after treatment.

b. $|t| = 1.0551$, df $= 4$

$0.9195 \leq 1.0551 \leq 1.4759$

$ t_{0.20} \leq 1.0551 \leq t_{0.10}$

$ 0.10 \leq p \leq 0.20$

10.57 a. The common characteristic that makes these data paired is the programmer.

b. H_0: $\mu_D = 0$; H_a: $\mu_D < 0$;

TS: $T = \frac{\overline{D} - \Delta_0}{S_D/\sqrt{n}}$; RR: $T \leq -3.5518$

$t = \frac{-6.44 - 0}{7.64/\sqrt{21}} = -3.8636 \leq -3.5518$

There is evidence to suggest the population mean runtime for Java programs is greater than the population mean runtime for C++ programs.

c. $|t| = 3.8638$, df $= 20$

$3.8495 \leq p \leq 4.5385$

$t_{0.0005} \leq p \leq t_{0.0001}$

$0.0001 \leq p \leq 0.0005$

10.59 a. $\overline{d} \pm t_{0.005} \frac{s_D}{\sqrt{n}}$

$= 2.80 \pm (2.9768)\frac{4.11}{\sqrt{15}}$

$= (-0.3584, 5.9584)$

b. Since 0 is included in this CI, there is no evidence to suggest that the pitching speed is, on average, faster from a higher mound.

10.61 a. H_0: $\mu_D = 0$; H_a: $\mu_D > 0$;

TS: $T = \frac{\overline{D} - \Delta_0}{S_D/\sqrt{n}}$; RR: $T \geq 4.0493$

$t = \frac{0.1799 - 0}{0.2096/\sqrt{46}} = 5.8213 \geq 4.0493$

There is evidence to suggest the population mean autofocus shutter lag is greater than the population mean prefocus shutter lag.

b. $|t| = 5.8213$, df $= 45$

$5.8123 \geq 4.0493 \Rightarrow 5.8123 \geq t_{0.0001} \Rightarrow p \leq 0.0001$

10.63 H_0: $\mu_D = 0$; H_a: $\mu_D > 0$;

TS: $T = \frac{\overline{D} - \Delta_0}{S_D/\sqrt{n}}$; RR: $T \geq 1.8331$

$t = \frac{42.4 - 0}{235.7/\sqrt{10}} = 0.5689$

There is no evidence to suggest the population mean salt content before potatoes is greater than the population mean salt content after potatoes.

10.65 a. H_0: $\mu_D = 0$; H_a: $\mu_D > 0$;

TS: $T = \frac{\overline{D} - \Delta_0}{S_D/\sqrt{n}}$; RR: $T \geq 1.8946$

$t = \frac{0.213 - 0}{1.271/\sqrt{8}} = 0.4740$

There is no evidence to suggest the population mean electromagnetic emission from the old antenna is greater than the population mean electromagnetic emission from the new antenna.

b. $|t| = 0.4740$, df $= 7$

$0.4740 \leq 0.8960 \Rightarrow 0.4740 \leq t_{0.20} \Rightarrow p \geq 0.20$

10.67 a. The common characteristic that makes these data paired is the exam score.

b. H_0: $\mu_D = 0$; H_a: $\mu_D \neq 0$;

TS: $T = \frac{\overline{D} - \Delta_0}{S_D/\sqrt{n}}$; RR: $|T| \geq 2.3646$

$t = \frac{0.588 - 0}{0.969/\sqrt{8}} = 1.7154$

There is no evidence to suggest the written-manual population mean assembly time is different from the interactive-video population mean assembly time.

10.69 a. The common characteristic that makes these data paired is the patient with high fever.

b. H_0: $\mu_D = 0$; H_a: $\mu_D > 0$;

TS: $T = \frac{\overline{D} - \Delta_0}{S_D/\sqrt{n}}$; RR: $T \geq 1.8331$

$t = \frac{2.86 - 0}{2.68/\sqrt{10}} = 3.3746 \geq 1.8331$

There is evidence to suggest the population mean temperature before the drug is greater than the population mean temperature after the drug.

c. $|t| = 3.3746$, df $= 9$

$3.2498 \leq p \leq 4.2968$

$t_{0.005} \leq p \leq t_{0.001}$

$0.001 \leq p \leq 0.005$

d. Since 9 of the 10 differences are positive and most of these are more than 1 degree of difference, we should expect the hypothesis test to be significant.

10.71 a. The common characteristic that makes these data paired is the force.

b. H_0: $\mu_D = 0$; H_a: $\mu_D < 0$;

TS: $T = \frac{\overline{D} - \Delta_0}{S_D/\sqrt{n}}$; RR: $T \leq -2.7638$

$t = \frac{-11.4-0}{6.57/\sqrt{11}} = -5.7564 \leq -2.7638$

There is evidence to suggest the old-formula population mean resilience is less than the new-formula population mean resilience.

c. Since all of the differences are negative, we should expect the hypothesis test to be significant.

Section 10.4: Comparing Two Population Proportions Using Large Samples

10.73

	Mean	Variance	Standard deviation	Probability
(a)	−0.020	0.000579	0.0241	0.0034
(b)	0.040	0.001751	0.0418	0.0280
(c)	0.010	0.001557	0.0395	0.7804
(d)	−0.070	0.000858	0.0293	0.8472
(e)	−0.120	0.000275	0.0166	0.9999
(f)	0.041	0.000914	0.0302	0.9527

10.75 a. TS: $Z = \dfrac{\widehat{P}_1 - \widehat{P}_2}{\sqrt{\widehat{P}_c(1-\widehat{P}_c)\left(\frac{1}{n_1}+\frac{1}{n_2}\right)}}$;

RR: $Z \geq 1.6449$

$z = \dfrac{0.80-0.7714}{\sqrt{(0.7854)(0.2146)\left(\frac{1}{525}+\frac{1}{405}\right)}} = 1.1137$

Do not reject H_0.

$p = \text{P}(Z \geq 1.1137) = 0.1327$

b. TS: $Z = \dfrac{\widehat{P}_1 - \widehat{P}_2}{\sqrt{\widehat{P}_c(1-\widehat{P}_c)\left(\frac{1}{n_1}+\frac{1}{n_2}\right)}}$; RR: $Z \leq -2.3263$

$z = \dfrac{0.4334-0.4853}{\sqrt{(0.46)(0.54)\left(\frac{1}{646}+\frac{1}{680}\right)}} = -1.8938$

Do not reject H_0.

$p = \text{P}(Z \leq -1.8938) = 0.0291$

c. TS: $Z = \dfrac{\widehat{P}_1 - \widehat{P}_2}{\sqrt{\widehat{P}_c(1-\widehat{P}_c)\left(\frac{1}{n_1}+\frac{1}{n_2}\right)}}$; RR: $|Z| \geq 2.2414$

$z = \dfrac{0.3176-0.4135}{\sqrt{(0.3666)(0.6334)\left(\frac{1}{255}+\frac{1}{266}\right)}} = -2.2705 \leq -2.2414$

Reject H_0.

$p = 2\text{P}(Z \leq -2.2414) = 0.0232$

d. TS: $Z = \dfrac{\widehat{P}_1 - \widehat{P}_2}{\sqrt{\widehat{P}_c(1-\widehat{P}_c)\left(\frac{1}{n_1}+\frac{1}{n_2}\right)}}$; RR: $|Z| \geq 3.2905$

$z = \dfrac{0.6299-0.6210}{\sqrt{(0.6252)(0.3748)\left(\frac{1}{1440}+\frac{1}{1562}\right)}} = 0.5012$

Do not reject H_0.

$p = 2\text{P}(Z \geq 0.5012) = 0.6163$

10.77 a. $(\widehat{p}_1 - \widehat{p}_2) \pm z_{0.025}\sqrt{\frac{\widehat{p}_1(1-\widehat{p}_1)}{n_1} + \frac{\widehat{p}_2(1-\widehat{p}_2)}{n_2}}$

$= (0.5928 - 0.6219)$

$\pm (1.96)\sqrt{\frac{(0.5928)(0.4072)}{388} + \frac{(0.6219)(0.3781)}{402}}$

$= (-0.0972, 0.0390)$

b. $(\widehat{p}_1 - \widehat{p}_2) \pm z_{0.025}\sqrt{\frac{\widehat{p}_1(1-\widehat{p}_1)}{n_1} + \frac{\widehat{p}_2(1-\widehat{p}_2)}{n_2}}$

$= (0.8996 - 0.9377)$

$\pm (1.96)\sqrt{\frac{(0.8996)(0.1004)}{528} + \frac{(0.9377)(0.0623)}{530}}$

$= (-0.0710, -0.0052)$

c. $(\widehat{p}_1 - \widehat{p}_2) \pm z_{0.005}\sqrt{\frac{\widehat{p}_1(1-\widehat{p}_1)}{n_1} + \frac{\widehat{p}_2(1-\widehat{p}_2)}{n_2}}$

$= (0.5111 - 0.5155)$

$\pm (2.5758)\sqrt{\frac{(0.5111)(0.4889)}{180} + \frac{(0.5155)(0.4845)}{194}}$

$= (-0.1376, 0.1289)$

d. $(\widehat{p}_1 - \widehat{p}_2) \pm z_{0.05}\sqrt{\frac{\widehat{p}_1(1-\widehat{p}_1)}{n_1} + \frac{\widehat{p}_2(1-\widehat{p}_2)}{n_2}}$

$= (0.7413 - 0.7030)$

$\pm (1.6449)\sqrt{\frac{(0.7413)(0.2587)}{2300} + \frac{(0.7030)(0.2970)}{2404}}$

$= (0.0168, 0.0598)$

10.79 $H_0: p_1 - p_2 = 0$; $H_a: p_1 - p_2 \neq 0$;

TS: $Z = \dfrac{\widehat{P}_1 - \widehat{P}_2}{\sqrt{\widehat{P}_c(1-\widehat{P}_c)\left(\frac{1}{n_1}+\frac{1}{n_2}\right)}}$; RR: $|Z| \geq 1.96$

$z = \dfrac{0.525-0.52}{\sqrt{(0.522)(0.478)\left(\frac{1}{200}+\frac{1}{550}\right)}} = 0.1523$

There is no evidence to suggest that the population proportin of repeat customers is different for Ford and Chevrolet.

10.81 a. $H_0: p_1 - p_2 = 0$; $H_a: p_1 - p_2 > 0$;

TS: $Z = \dfrac{\widehat{P}_1 - \widehat{P}_2}{\sqrt{\widehat{P}_c(1-\widehat{P}_c)\left(\frac{1}{n_1}+\frac{1}{n_2}\right)}}$; RR: $Z \geq 3.0902$

$z = \dfrac{0.686-0.571}{\sqrt{(0.625)(0.375)\left(\frac{1}{347}+\frac{1}{387}\right)}} = 3.2086 \geq 3.0902$

There is evidence to suggest the population proportion of 18 to 29-year-olds who believe movies are getting better is greater than the population proportion of 30 to 49-year-olds who believe movies are getting better.

b. $p = \text{P}(Z \geq 3.2086) = 0.000667$

10.83 a. $n_1\widehat{p}_1 = 530 \geq 5$, $n_1(1 - \widehat{p}_1) = 525 \geq 5$, $n_2\widehat{p}_2 = 825 \geq 5$, $n_2(1 - \widehat{p}_2) = 838 \geq 5$

b. $H_0: p_1 - p_2 = 0$; $H_a: p_1 - p_2 > 0$;

TS: $Z = \dfrac{\widehat{P}_1 - \widehat{P}_2}{\sqrt{\widehat{P}_c(1-\widehat{P}_c)\left(\frac{1}{n_1}+\frac{1}{n_2}\right)}}$; RR: $Z \geq 2.3263$

$z = \dfrac{0.502-0.496}{\sqrt{(0.4985)(0.5015)\left(\frac{1}{1055}+\frac{1}{1663}\right)}} = 0.3190$

There is no evidence to suggest the population proportion of carpoolers crossing the George Washington Bridge is greater than the population proportion of carpoolers using the Lincoln Tunnel.

c. $p = P(Z \geq 0.3190) = 0.3749$

10.85 a. $\widehat{p}_1 = \frac{71}{255} = 0.2784$, $\widehat{p}_2 = \frac{55}{237} = 0.2321$

b. $H_0: p_1 - p_2 = 0$; $H_a: p_1 - p_2 > \neq 0$;

TS: $Z = \dfrac{\widehat{P}_1 - \widehat{P}_2}{\sqrt{\widehat{P}_c(1-\widehat{P}_c)\left(\frac{1}{n_1}+\frac{1}{n_2}\right)}}$; RR: $|Z| \geq 2.5758$

$z = \dfrac{0.2784 - 0.2321}{\sqrt{(0.256)(0.744)\left(\frac{1}{255}+\frac{1}{237}\right)}} = 1.1773$

There is no evidence to suggest the population proportion of people who experience relief due to the antihistamine is different from the population proportion of people who experience relief from butterbur extract.

10.87 a. $\widehat{p}_1 = \frac{8}{106} = 0.0755$, $\widehat{p}_2 = \frac{12}{121} = 0.0992$

b. $n_1 \widehat{p}_1 = 8 \geq 5$, $n_1(1-\widehat{p}_1) = 98 \geq 5$
$n_2 \widehat{p}_2 = 12 \geq 5$, $n_2(1-\widehat{p}_2) = 109 \geq 5$

c. $H_0: p_1 - p_2 = 0$; $H_a: p_1 - p_2 \neq 0$;

TS: $Z = \dfrac{\widehat{P}_1 - \widehat{P}_2}{\sqrt{\widehat{P}_c(1-\widehat{P}_c)\left(\frac{1}{n_1}+\frac{1}{n_2}\right)}}$; RR: $|Z| \geq 1.96$

$z = \dfrac{0.0755 - 0.0992}{\sqrt{(0.0881)(0.9119)\left(\frac{1}{106}+\frac{1}{121}\right)}} = -0.6286$

There is no evidence to suggest the population proportion of defective lenses is different for Process A than Process B.

10.89 $H_0: p_1 - p_2 = 0$; $H_a: p_1 - p_2 \neq 0$;

TS: $Z = \dfrac{\widehat{P}_1 - \widehat{P}_2}{\sqrt{\widehat{P}_c(1-\widehat{P}_c)\left(\frac{1}{n_1}+\frac{1}{n_2}\right)}}$; RR: $|Z| \geq 1.96$

$z = \dfrac{0.1808 - 0.1822}{\sqrt{(0.1816)(0.8185)\left(\frac{1}{625}+\frac{1}{730}\right)}} = -0.0663$

There is no evidence to suggest the population proportion of veterans living in Colorado Springs and Jacksonville is different.

10.91 a. $H_0: p_1 - p_2 = 0$; $H_a: p_1 - p_2 < 0$;

TS: $Z = \dfrac{\widehat{P}_1 - \widehat{P}_2}{\sqrt{\widehat{P}_c(1-\widehat{P}_c)\left(\frac{1}{n_1}+\frac{1}{n_2}\right)}}$; RR: $Z \leq -1.6449$

$z = \dfrac{0.181 - 0.26}{\sqrt{(0.216)(0.784)\left(\frac{1}{570}+\frac{1}{120}\right)}} = -3.0677 \leq -1.6449$

There is evidence to suggest the population proportion of 18- to 29-year-olds who obtain news every day from nightly network news shows is less than the population proportion of 30- to 49-year-olds who obtain news every day from night network news shows.

$p = P(Z \leq -3.0677) = 0.0011$

b. $H_0: p_1 - p_2 = 0$; $H_a: p_1 - p_2 < 0$;

TS: $Z = \dfrac{\widehat{P}_1 - \widehat{P}_2}{\sqrt{\widehat{P}_c(1-\widehat{P}_c)\left(\frac{1}{n_1}+\frac{1}{n_2}\right)}}$; RR: $Z \leq -2.3263$

$z = \dfrac{0.24 - 0.35}{\sqrt{(0.2990)(0.7010)\left(\frac{1}{450}+\frac{1}{520}\right)}} = -3.7319 \leq -2.3263$

There is evidence to suggest the population proportion of 18- to 29-year-olds who obtain news every day from nightly cable news shows is less than the population proportion of 30- to 49-year-olds who obtain news every day from night cable news shows.

$p = P(Z \leq -3.7319) = 0.000095$

c. $H_0: p_1 - p_2 = 0$; $H_a: p_1 - p_2 < 0$;

TS: $Z = \dfrac{\widehat{P}_1 - \widehat{P}_2}{\sqrt{\widehat{P}_c(1-\widehat{P}_c)\left(\frac{1}{n_1}+\frac{1}{n_2}\right)}}$; RR: $Z \leq -3.2905$

$z = \dfrac{0.361 - 0.421}{\sqrt{(0.3914)(0.6086)\left(\frac{1}{568}+\frac{1}{546}\right)}} = -2.0501$

There is no evidence to suggest the population proportion of 18- to 29-year-olds who obtain news every day from the Internet is less than the population proportion of 30- to 49-year-olds who obtain news every day from the Internet.

$p = P(Z \leq -2.0501) = 0.0202$

10.93 $H_0: p_1 - p_2 = 010$; $H_a: p_1 - p_2 > 0.10$;

TS: $Z = \dfrac{(\widehat{P}_1 - \widehat{P}_2) - 0.05}{\sqrt{\frac{\widehat{P}_1(1-\widehat{P}_1)}{n_1} + \frac{\widehat{P}_2(1-\widehat{P}_2)}{n_2}}}$;

RR: $Z \geq 2.3263$

$z = \dfrac{(0.3448 - 0.2145) - 0.10}{\sqrt{\frac{(0.3448)(0.6552)}{261} + \frac{(0.2145)(0.7855)}{303}}} = 0.8038$

There is no evidence to suggest the population proportion of homeowners planning a landscaping project is more than 0.10 greater than the population proportion of condominium owners planning a landscaping project.

Section 10.5: Comparing Two Population Variances or Standard Deviations

10.95 a. $F_{0.01} = 2.20$ **b.** $F_{0.01} = 3.15$
c. $F_{0.05} = 3.58$ **d.** $F_{0.001} = 4.99$ **e.** $F_{0.99} = 0.23$
f. $F_{0.99} = 0.12$ **g.** $F_{0.95} = 0.34$ **h.** $F_{0.999} = 0.25$

10.97 a. $H_0: \sigma_1^2 = \sigma_2^2$; $H_a: \sigma_1^2 > \sigma_2^2$;
TS: $F = S_1^2 / S_2^2$; RR: $F \geq 1.94$

b. $f = 44.89/17.64 = 2.5448 \geq 1.94$

There is evidence to suggest population variance 1 is greater than population variance 2.

c. df 30 and 24

$1.94 \leq 2.5448 \leq 2.58$
$F_{0.05} \leq \quad f \quad \leq F_{0.01}$
$0.01 \leq \quad p \quad \leq 0.05$

p value illustration:

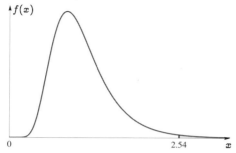

10.99 a. H_0: $\sigma_1^2 = \sigma_2^2$; H_a: $\sigma_1^2 \neq \sigma_2^2$;
TS: $F = S_1^2/S_2^2$; RR: $F \leq 0.17$ or $F \geq 4.54$

b. $f = 426.42/88.36 = 4.8259 \geq 4.54$

There is evidence to suggest that population variance 1 is different from population variance 2.

c. df 9 and 15

$3.89 \leq 4.8259 \leq 6.26$
$F_{0.01} \leq \quad f \quad \leq F_{0.001}$
$0.001 \leq p/2 \leq 0.01$
$0.002 \leq \quad p \quad \leq 0.02$

10.101 a. $\left(\frac{s_1^2}{s_2^2}\frac{1}{F_{0.05}}, \frac{s_1^2}{s_2^2}F_{0.05}\right)$

$= \left(\frac{17.2}{15.6}\frac{1}{3.39}, \frac{17.2}{15.6}(3.23)\right)$
$= (0.3254, 3.5608)$

b. $\left(\frac{s_1^2}{s_2^2}\frac{1}{F_{0.01}}, \frac{s_1^2}{s_2^2}F_{0.01}\right)$

$= \left(\frac{54.1}{32.6}\frac{1}{3.52}, \frac{54.1}{32.6}(3.52)\right)$
$= (0.4712, 5.8451)$

c. $\left(\frac{s_1^2}{s_2^2}\frac{1}{F_{0.01}}, \frac{s_1^2}{s_2^2}F_{0.01}\right)$

$= \left(\frac{3.35}{4.59}\frac{1}{2.70}, \frac{3.35}{4.59}(3.21)\right)$
$= (0.2703, 2.3458)$

d. $\left(\frac{s_1^2}{s_2^2}\frac{1}{F_{0.001}}, \frac{s_1^2}{s_2^2}F_{0.001}\right)$

$= \left(\frac{126.8}{155.3}\frac{1}{2.87}, \frac{126.8}{155.3}(3.07)\right)$
$= (0.2843, 2.5079)$

10.103 H_0: $\sigma_1^2 = \sigma_2^2$; H_a: $\sigma_1^2 > \sigma_2^2$;
TS: $F = S_1^2/S_2^2$; RR: $F \geq 2.39$

$f = 243.36/51.84 = 4.6944 \geq 2.39$

There is evidence to suggest the population variance in aerosol light absorption coefficient is greater in Africa than in South America.

10.105 H_0: $\sigma_1^2 = \sigma_2^2$; H_a: $\sigma_1^2 > \sigma_2^2$;
TS: $F = S_1^2/S_2^2$; RR: $F \geq 3.52$

$f = 37.45/8.004 = 4.6786 \geq 3.52$

There is evidence to suggest the population variance

in the weight of frozen turkeys from North Carolina is greater than the population variance in the weight of frozen turkeys from Minnesota.

10.107 a. H_0: $\sigma_1^2 = \sigma_2^2$; H_a: $\sigma_1^2 > \sigma_2^2$;
TS: $F = S_1^2/S_2^2$; RR: $F \geq 2.60$

$f = 110.25/38.44 = 2.8681 \geq 2.6041$

There is evidence to suggest the population variance in winning times for an ordinary race is greater than the population variance in winning times for a stakes race.

b. $\left(\frac{s_1^2}{s_2^2}\frac{1}{F_{0.01}}, \frac{s_1^2}{s_2^2}F_{0.01}\right)$

$= \left(\frac{110.25}{38.44}\frac{1}{2.60}, \frac{110.25}{38.44}(2.60)\right)$
$= (1.1014, 7.4689)$

10.109 a. H_0: $\sigma_1^2 = \sigma_2^2$; H_a: $\sigma_1^2 \neq \sigma_2^2$;
TS: $F = S_1^2/S_2^2$; RR: $F \leq 0.37$ or $F \geq 2.67$

$f = 550.70/991.95 = 0.5552$

There is no evidence to suggest the population variance in flight-delay times for Delta is different from the population variance in flight-delay times for United.

b. The normality assumption is probably not reasonable because the distributions are skewed right.

10.111 H_0: $\sigma_1^2 = \sigma_2^2$; H_a: $\sigma_1^2 > \sigma_2^2$;
TS: $F = S_1^2/S_2^2$; RR: $F \geq 3.29$

$f = 2284.827/254.867 = 8.9648 \geq 3.2940$

There is evidence to suggest the population variance in lodging per diem is greater in the Northeast than in the West.

10.113 a. H_0: $\sigma_1^2 = \sigma_2^2$; H_a: $\sigma_1^2 \neq \sigma_2^2$;
TS: $F = S_1^2/S_2^2$; RR: $F \leq 0.42$ or $F \geq 2.54$

$f = 9722.3^2/13119^2 = 0.5492$

There is no evidence to suggest the population variance in circulation for the *Sun-Times* is different from the population variance in circulation for the *Globe*.

b. The normality assumption is probably not reasonable because the circulation distributions could be skewed right.

10.115 a. H_0: $\sigma_1^2 = \sigma_2^2$; H_a: $\sigma_1^2 \neq \sigma_2^2$;
TS: $F = S_1^2/S_2^2$; RR: $F \leq 0.31$ or $F \geq 3.53$

$f = 7.84/2.89 = 2.7128$

There is no evidence to suggest the population variance in saccharin amount for Fishing Creek is different from the population variance in saccharin amount for Honest Tea.

b. df 19 and 14

$$2.40 \leq 2.7128 \leq 3.53$$
$$F_{0.05} \leq \quad f \quad \leq F_{0.01}$$
$$0.01 \leq \quad p/2 \quad \leq 0.05$$
$$0.02 \leq \quad p \quad \leq 0.10$$

Chapter Exercises

10.117 H_0: $\mu_1 - \mu_2 = 0$; H_a: $\mu_1 - \mu_2 \neq 0$;

TS: $Z = \dfrac{(\overline{X}_1 - \overline{X}_2) - 0}{\sqrt{\frac{\sigma_1^2}{n_1} + \frac{\sigma_2^2}{n_2}}}$; RR: $|Z| \geq 2.5758$

$z = \dfrac{(835.6 - 884.2) - 0}{\sqrt{\frac{3192.25}{22} + \frac{3956.41}{25}}} = -2.7903 \leq -2.5758$

There is evidence to suggest the population mean amount of corrosive material carried by trucks in North Carolina is different from the population meant amount of corrosive material carried by trucks in Virginia.

10.119 a. $\widehat{p}_1 = \frac{335}{544} = 0.6158$, $\widehat{p}_2 = \frac{381}{603} = 0.6318$

$n_1 \widehat{p}_1 = 335 \geq 5$, $n_1(1 - \widehat{p}_1) = 209 \geq 5$,
$n_2 \widehat{p}_2 = 381 \geq 5$, $n_2(1 - \widehat{p}_2) = 222 \geq 5$

b. H_0: $p_1 - p_2 = 0$; H_a: $p_1 - p_2 \neq 0$; TS:
$Z = \dfrac{\widehat{P}_1 - \widehat{P}_2}{\sqrt{\widehat{P}_c(1 - \widehat{P}_c)\left(\frac{1}{n_1} + \frac{1}{n_2}\right)}}$; RR: $|Z| \geq 2.5758$

$z = z = \dfrac{0.6158 - 0.6318}{\sqrt{(0.6242)(0.3758)\left(\frac{1}{544} + \frac{1}{603}\right)}} = -0.5598$

There is no evidence to suggest the two population proportions are different.

c. $p = 2P(Z \leq -0.5598) = 0.5756$

10.121 a. H_0: $p_1 - p_2 = 0$; H_a: $p_1 - p_2 > 0$; TS:
$Z = \dfrac{\widehat{P}_1 - \widehat{P}_2}{\sqrt{\widehat{P}_c(1 - \widehat{P}_c)\left(\frac{1}{n_1} + \frac{1}{n_2}\right)}}$; RR: $Z \geq 2.3263$

$z = \dfrac{0.77 - 0.749}{\sqrt{(0.7593)(0.2407)\left(\frac{1}{909} + \frac{1}{923}\right)}} = 1.0728$

There is no evidence to suggest the population proportion of residents in Ohio who recycle newspapers is greater than the population proportion of residents in Florida.

b. $p = P(Z \geq 1.0728) = 0.1417$

10.123 H_0: $\sigma_1^2 = \sigma_2^2$; H_a: $\sigma_1^2 > \sigma_2^2$;
TS: $F = S_1^2/S_2^2$; RR: $F \geq 5.06$

$f = 0.0196/0.0025 = 7.84 \geq 5.06$

There is evidence to suggest the population variance in aluminum fuselage thickness is greater than the population variance in carbon-fiber fuselage thickness.

10.125 H_0: $\mu_1 - \mu_2 = 0$; H_a: $\mu_1 - \mu_2 > 0$;

TS: $T' = \dfrac{(\overline{X}_1 - \overline{X}_2) - 0}{\sqrt{\left(\frac{S_1^2}{n_1} + \frac{S_2^2}{n_2}\right)}}$; RR: $T' \geq 3.7874$

$t' = \dfrac{(1.08 - 0.58) - 0}{\sqrt{\frac{0.32^2}{10} + \frac{0.52^2}{10}}} = 2.5896$

There is no evidence to suggest the population mean PCB level in Smallmouth Bass at Bull's Bridge is greater than the population mean PCB level in Smallmouth Bass at Lake Zoar.

10.127 H_0: $\mu_1 - \mu_2 = 0$; H_a: $\mu_1 - \mu_2 > 0$;

TS: $T = \dfrac{(\overline{X}_1 - \overline{X}_2) - 0}{\sqrt{S_p^2\left(\frac{1}{n_1} + \frac{1}{n_2}\right)}}$; RR: $T \geq 3.5518$

$t = \dfrac{(325.5 - 267.9) - 0}{\sqrt{(60.2795)^2\left(\frac{1}{12} + \frac{1}{10}\right)}} = 2.2317$

There is no evidence to suggest the population mean Rolling Stones concert ticket price is greater than the population mean Coldplay concert ticket price.

10.129 H_0: $\mu_1 - \mu_2 = 0$; H_a: $\mu_1 - \mu_2 \neq 0$;

TS: $T = \dfrac{(\overline{X}_1 - \overline{X}_2) - 0}{\sqrt{S_p^2\left(\frac{1}{n_1} + \frac{1}{n_2}\right)}}$; RR: $|T| \geq 2.0017$

$t = \dfrac{(13.938 - 20.0) - 0}{\sqrt{(8.5402)^2\left(\frac{1}{25} + \frac{1}{35}\right)}} = -2.6898 \leq -2.0017$

There is evidence to suggest the population mean fine particulate measure is different in these two areas.

10.131 a. H_0: $\mu_1 - \mu_2 = 0$; H_a: $\mu_1 - \mu_2 \neq 0$

TS: $T = \dfrac{(\overline{X}_1 - \overline{X}_2) - 0}{\sqrt{S_p^2\left(\frac{1}{n_1} + \frac{1}{n_2}\right)}}$; RR: $|T| \geq 2.6778$

$t = \dfrac{(75.1 - 80.9) - 0}{\sqrt{(2.405)^2\left(\frac{1}{26} + \frac{1}{26}\right)}} = -8.6946 \leq -2.6778$

There is evidence to suggest the population mean number of yearly pro bono hours is different at these two law firms.

b. $(\overline{x}_1 - \overline{x}_2) \pm t_{0.005}\sqrt{s_p^2\left(\frac{1}{n_1} + \frac{1}{n_2}\right)}$

$= (75.1 - 80.9) \pm (2.6778)\sqrt{(2.405)^2\left(\frac{1}{26} + \frac{1}{26}\right)}$

$= (-7.5863, -4.0137)$

c. This confidence interval supports our conclusion in part (a) since the CI includes 0 and is entirely below 0.

10.133 For each hypothesis test, use

H_0: $\mu_1 - \mu_2 = 0$; H_a: $\mu_1 - \mu_2 \neq 0$;

TS: $T = \dfrac{(\overline{X}_1 - \overline{X}_2) - 0}{\sqrt{S_p^2\left(\frac{1}{n_1} + \frac{1}{n_2}\right)}}$

Race versus age:

$t = \dfrac{(120.9 - 268.0) - 0}{\sqrt{(194.756)^2\left(\frac{1}{15} + \frac{1}{15}\right)}} = -2.0685$

$p = 2P(T \leq -2.0685) = 0.0479$

Reject H_0.

Race versus disability:
$$t = \frac{(120.9 - 175.0) - 0}{\sqrt{(199.589)^2\left(\frac{1}{15} + \frac{1}{16}\right)}} = -0.7547$$
$$p = 2P(T \leq -0.7547) = 0.4565$$

Do not reject H_0.

Age versus disability:
$$t = \frac{(268.0 - 175.0) - 0}{\sqrt{(190.838)^2\left(\frac{1}{15} + \frac{1}{16}\right)}} = 1.3559$$
$$p = 2P(T \geq 1.3559) = 0.1856$$

Do not reject H_0.

Chapter 11: The Analysis of Variance

Section 11.1: One-Way ANOVA

11.1 a. $n_1 = 9$, $n_2 = 7$, $n_3 = 6$

b. $t_1 = 260$, $t_2 = 229$, $t_3 = 195$

c. $\sum_{i=1}^{3} \sum_{j=1}^{n_i} x_{ij}^2 = 33^3 + 27^2 + \cdots + 28^2 = 21{,}602$

11.3 a. $t_1 = 775$, $t_2 = 745$, $t_3 = 768$, $t_4 = 753$

b. $\sum_{i=1}^{4} \sum_{j=1}^{n_i} x_{ij}^2 = 162^2 + \cdots + 157^2 = 465{,}193$

c. SST $= 2808.95$, SSA $= 112.55$, SSE $= 2696.40$

d. MSA $= 112.55/3 = 37.5167$,
MSE $= 2696.40/16 = 168.525$

e. $f = 37.5167/168.525 = 0.2226$

11.5
ANOVA Summary Table

Source of variation	Sum of squares	Degrees of freedom	Mean square	F	p value
Factor	584.1	4	146.0250	0.57	0.6864
Error	12,062.1	47	256.6404		
Total	12,646.2	51			

a. H_0: $\mu_1 = \mu_2 = \mu_3 = \mu_4 = \mu_5$;
H_a: $\mu_i \neq \mu_j$ for some $i \neq j$

b. df 4 and 47. RR: $F \geq 2.57$

c. $f = 146.0250/256.6404 = 0.57$

There is no evidence to suggest at least two of the population means are different.

11.7 a. H_0: $\mu_1 = \mu_2 = \mu_3 = \mu_4$;
H_a: $\mu_i \neq \mu_j$ for some $i \neq j$;
TS: $F = $ MSA/MSE; RR: $F \geq 3.10$

b.
ANOVA Summary Table

Source of variation	Sum of squares	Degrees of freedom	Mean square	F	p value
Factor	67000.33	3	22333.44	12.39	0.0001
Error	36039.67	20	1801.98		
Total	103040.00	23			

c. $f = 22333.44/1801.98 = 12.39 \geq 3.10$

There is evidence to suggest at least two of the population means are different.

11.9 H_0: $\mu_1 = \mu_2 = \mu_3 = \mu_4 = \mu_5$;
H_a: $\mu_i \neq \mu_j$ for some $i \neq j$;
TS: $F = $ MSA/MSE; RR: $F \geq 2.48$
SST $= 80133 - 2531^2/86 = 5645.0814$

SSA $= 75991.9673 - 2531^2/86 = 1504.0487$
SSE $= 5645.0814 - 1504.0487 = 4141.0327$

MSA $= 1504.0487/4 = 376.0122$
MSE $= 4141.0327/81 = 51.1239$

$f = 376.0122/51.1239 = 7.35 \geq 2.48$

There is evidence to suggest at least two population mean times are different.

11.11 H_0: $\mu_1 = \mu_2 = \mu_3$;
H_a: $\mu_i \neq \mu_j$ for some $i \neq j$;
TS: $F = $ MSA/MSE; RR: $F \geq 5.19$

SST $= 34.9466 - 32.8778 = 2.0688$
SSA $= 33.3857 - 32.8778 = 0.5079$
SSE $= 2.0688 - 0.5079 = 1.5609$

MSA $= 0.5079/2 = 0.2540$
MSE $= 1.5609/39 = 0.0400$

$f = 0.2540/0.0400 = 6.35 \geq 5.19$

There is evidence to suggest at least two of the population mean weights are different.

11.13 H_0: $\mu_1 = \mu_2 = \mu_3 = \mu_4$;
H_a: $\mu_i \neq \mu_j$ for some $i \neq j$;
TS: $F = $ MSA/MSE; RR: $F \geq 4.13$

SST $= 4186.71 - 493.3^2/64 = 384.4461$
SSA $= 3900.0179 - 493.3^2/64 = 97.7540$
SSE $= 384.4461 - 97.7540 = 286.6921$

MSA $= 97.7540/3 = 32.5847$
MSE $= 286.6921/60 = 4.7782$

$f = 32.5847/4.7782 = 6.82 \geq 4.13$

There is evidence to suggest at least two of the population mean checkpoint times are different.

11.15 H_0: $\mu_1 = \mu_2 = \mu_3$;
H_a: $\mu_i \neq \mu_j$ for some $i \neq j$;
TS: $F = $ MSA/MSE; RR: $F \geq 3.16$

SST $= 198033.0 - 168222.15 = 29810.85$
SSA $= 173574.85 - 168222.15 = 5352.70$
SSE $= 29810.85 - 5352.70 = 24458.15$
MSA $= 5352.70/2 = 2676.35$
MSE $= 24458.15/57 = 429.0904$

$f = 2676.35/429.0904 = 6.24 \geq 3.16$

There is evidence to suggest at least two population mean numbers of plants per seized plot are different.

11.17 H_0: $\mu_1 = \mu_2 = \mu_3 = \mu_4$;
H_a: $\mu_i \neq \mu_j$ for some $i \neq j$;
TS: $F = $ MSA/MSE; RR: $F \geq 3.10$

SST $= 52951.0 - 51245.0417 = 1705.9583$
SSA $= 51429.1667 - 51245.0417 = 184.1250$

SSE $= 1705.9583 - 184.1250 = 1521.8333$

MSA $= 184.1250/3 = 61.3750$

MSE $= 1521.8333/20 = 76.0917$

$f = 61.3750/76.0917 = 0.81$

There is no evidence to suggest at least two population mean thaw depths are different.

11.19 $H_0: \mu_1 = \mu_2 = \mu_3 = \mu_4$;
$H_a: \mu_i \neq \mu_j$ for some $i \neq j$;
TS: $F = $ MSA/MSE; RR: $F \geq 2.72$

SST $= 2164.0 - 1920.8 = 243.2$

SSA $= 1930.0 - 1920.8 = 9.2$

SSE $= 243.2 - 9.2 = 234.0$

MSA $= 9.2/3 = 3.0667$

MSE $= 234.0/76 = 3.0789$

$f = 3.0667/3.0789 = 1.00$

There is no evidence to suggest at least two population mean numbers of unhealthy days per 30-day period are different.

11.21 a. $H_0: \mu_1 = \mu_2 = \mu_3 = \mu_4$;
$H_a: \mu_i \neq \mu_j$ for some $i \neq j$;
TS: $F = $ MSA/MSE; RR: $F \geq 3.01$

SST $= 160,502,179.0 - 158,474,706.04 = 2,027,472.96$

SSA $= 159,079,851.57 - 158,474,706.04 = 605,145.53$

SSE $= 2,027,472.96 - 605,145.53 = 1,422,327.43$

MSA $= 605,145.53/3 = 201,715.18$

MSE $= 1,422,327.43/24 = 59,263.64$

$f = 201,715.18/59,263.64 = 3.40 \geq 3.01$

There is evidence to suggest at least two of the population mean pressures are different.

b. Recommend Holder since this broom has the highest mean pressure.

11.23 a. $H_0: \mu_1 = \mu_2 = \mu_3 = \mu_4 = \mu_5$;
$H_a: \mu_i \neq \mu_j$ for some $i \neq j$;
TS: $F = $ MSA/MSE; RR: $F \geq 5.36$

SST $= 4678.87 - 533.3^2/61 = 16.4292$

SSA $= 4672.4060 - 533.3^2/61 = 9.9652$

SSE $= 16.4292 - 9.9652 = 6.4640$

MSA $= 9.9652/4 = 2.4913$

MSE $= 6.4640/56 = 0.1154$

$f = 2.4913/0.1154 = 21.58 \geq 5.36$

There is evidence to suggest at least two population mean weights are different.

b. Buy at Weis since these bags have the largest sample mean weight.

11.25 Answers will vary.

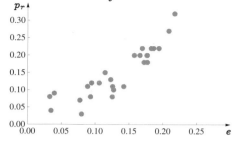

As effect size increases, the probability of rejecting H_0 also increases. This graph illustrates the power of the test, the probability of rejecting the null hypothesis for a specific alternative.

Section 11.2: Isolating Differences

11.27 a. df 3 and 18, $Q_{0.05} = 3.609$

b. df 4 and 40, $Q_{0.05} = 3.791$

c. df 4 and 50, $Q_{0.01} = 4.634$

d. df 5 and 50, $Q_{0.01} = 4.863$

e. df 6 and 30, $Q_{0.001} = 6.469$

11.29 a.

$\bar{x}_{1\cdot}$	$\bar{x}_{2\cdot}$	$\bar{x}_{3\cdot}$	$\bar{x}_{4\cdot}$
-33.44	-14.83	0.48	4.30

b.

$\bar{x}_{3\cdot}$	$\bar{x}_{2\cdot}$	$\bar{x}_{4\cdot}$	$\bar{x}_{1\cdot}$
1.30	1.41	1.50	1.62

c.

$\bar{x}_{4\cdot}$	$\bar{x}_{2\cdot}$	$\bar{x}_{3\cdot}$	$\bar{x}_{1\cdot}$	$\bar{x}_{5\cdot}$
51.92	54.21	60.80	64.35	64.85

11.31 $c = \binom{4}{2} = 6$, df $= 80 - 4 = 76$
$t_{0.01/12} = t_{0.0008} = 3.2603$

Difference	Bonferroni CI	Significantly different
$\mu_1 - \mu_2$	$(-7.62, 3.64)$	No
$\mu_1 - \mu_3$	$(-2.54, 8.72)$	No
$\mu_1 - \mu_4$	$(-11.80, -0.54)$	Yes
$\mu_2 - \mu_3$	$(-0.55, 10.71)$	No
$\mu_2 - \mu_4$	$(-9.81, 1.45)$	No
$\mu_3 - \mu_4$	$(-14.89, -3.63)$	Yes

11.33 $c = \binom{3}{2} = 3$, df $= 67 - 3 = 64$
$t_{0.05/6} = t_{0.0083} = 2.4585$

Difference	Bonferroni CI	Significantly different
$\mu_1 - \mu_2$	(0.06, 1.28)	Yes
$\mu_1 - \mu_3$	(−1.02, 0.16)	No
$\mu_2 - \mu_3$	(−1.73, −0.47)	Yes

11.35 a. df 3 and 20

$4.94 \leq 7.69 \leq 9.55$

$F_{0.01} \leq f \leq F_{0.001}$

$0.001 \leq p \leq 0.01$

There is evidence to suggest at least two population means are different.

b. $c = \binom{4}{2} = 6$, df $= 24 - 4 = 20$

$t_{0.01/12} = t_{0.00083} = 3.6303$

Difference	Bonferroni CI	Significantly different
$\mu_1 - \mu_2$	(0.44, 7.35)	Yes
$\mu_1 - \mu_3$	(−3.56, 3.36)	No
$\mu_1 - \mu_4$	(−1.86, 5.06)	No
$\mu_2 - \mu_3$	(−7.45, −0.54)	Yes
$\mu_2 - \mu_4$	(−5.76, 1.16)	No
$\mu_3 - \mu_4$	(−1.76, 5.16)	No

11.37 a. df 3 and 40, $9.10 > 5.528$

$\Rightarrow f > F_{0.001} \Rightarrow p < 0.0001$

b. $c = \binom{4}{2} = 6$, df $= 44 - 4 = 40$

$t_{0.05/12} = t_{0.0042} = 2.7759$

Difference	Bonferroni CI	Significantly different
$\mu_1 - \mu_2$	(0.87, 1.25)	Yes
$\mu_1 - \mu_3$	(0.56, 0.95)	Yes
$\mu_1 - \mu_4$	(0.20, 0.59)	Yes
$\mu_2 - \mu_3$	(−0.49, −0.12)	Yes
$\mu_2 - \mu_4$	(−0.85, −0.48)	Yes
$\mu_3 - \mu_4$	(−0.55, −0.17)	Yes

Graph of results:

$\overline{x}_2.$	$\overline{x}_3.$	$\overline{x}_4.$	$\overline{x}_1.$
0.2567	0.5601	0.9206	1.3164

There are no horizontal lines under any sample means since the Bonferroni CIs suggest all population means are different.

11.39 a. H_0: $\mu_1 = \mu_2 = \mu_3 = \mu_4$;

H_a: $\mu_i \neq \mu_j$ for some $i \neq j$;

TS: $F = \text{MSA/MSE}$; RR: $F \geq 3.10$

$f = 207.2161/20.6498 = 10.03 \geq 3.10$

There is evidence to suggest at least two population mean honey production amounts are different.

b. $c = \binom{4}{2} = 6$, df $= 24 - 4 = 20$

$t_{0.05/12} = t_{0.0042} = 2.9271$

Difference	Bonferroni CI	Significantly different
$\mu_1 - \mu_2$	(−15.05, 0.31)	No
$\mu_1 - \mu_3$	(−6.53, 8.83)	No
$\mu_1 - \mu_4$	(−18.76, −3.40)	Yes
$\mu_2 - \mu_3$	(0.84, 16.20)	Yes
$\mu_2 - \mu_4$	(−11.40, 3.96)	No
$\mu_3 - \mu_4$	(−19.91, −4.55)	Yes

c. Graph of results:

$\overline{x}_3.$	$\overline{x}_1.$	$\overline{x}_2.$	$\overline{x}_4.$
65.55	66.70	74.07	77.78

11.41 a. H_0: $\mu_1 = \mu_2 = \mu_3$;

H_a: $\mu_i \neq \mu_j$ for some $i \neq j$;

TS: $F = \text{MSA/MSE}$; RR: $F \geq 3.32$

$f = 6524.7603/211.4979 = 30.85 \geq 3.32$

There is evidence to suggest at least two population mean stun gun voltages are different.

b. df 3 and 30, $Q_{0.05} = 3.486$

Difference	Tukey CI	Significantly different
$\mu_1 - \mu_2$	(32.86, 63.56)	Yes
$\mu_1 - \mu_3$	(17.75, 49.08)	Yes
$\mu_2 - \mu_3$	(−29.76, 0.17)	No

c. Graph of results:

$\overline{x}_2.$	$\overline{x}_3.$	$\overline{x}_1.$
67.80	82.59	116.01

11.43 a. H_0: $\mu_1 = \mu_2 = \mu_3 = \mu_4 = \mu_5$;

H_a: $\mu_i \neq \mu_j$ for some $i \neq j$;

TS: $F = \text{MSA/MSE}$; RR: $F \geq 4.43$

$f = 21.7904/1.6996 = 12.82 \geq 4.43$

There is evidence to suggest at least two population mean distances between courts are different.

b. $c = \binom{5}{2} = 10$, df $= 25 - 5 = 20$

$t_{0.05/20} = t_{0.0025} = 3.1534$

Difference	Bonferroni CI	Significantly different
$\mu_1 - \mu_2$	$(-4.66,\ 0.54)$	No
$\mu_1 - \mu_3$	$(-3.36,\ 1.84)$	No
$\mu_1 - \mu_4$	$(-4.36,\ 0.84)$	No
$\mu_1 - \mu_5$	$(-8.04, -2.84)$	Yes
$\mu_2 - \mu_3$	$(-1.30,\ 3.90)$	No
$\mu_2 - \mu_4$	$(-2.30,\ 2.90)$	No
$\mu_2 - \mu_5$	$(-5.98, -0.78)$	Yes
$\mu_3 - \mu_4$	$(-3.60,\ 1.60)$	No
$\mu_3 - \mu_5$	$(-7.28, -2.08)$	Yes
$\mu_4 - \mu_5$	$(-6.28, -1.08)$	Yes

c. Graph of results:

$\bar{x}_{1.}$	$\bar{x}_{3.}$	$\bar{x}_{4.}$	$\bar{x}_{2.}$	$\bar{x}_{5.}$
14.28	15.04	16.04	16.34	19.72

11.45 a. $H_0: \mu_1 = \mu_2 = \mu_3 = \mu_4$;
$H_a: \mu_i \neq \mu_j$ for some $i \neq j$;
TS: $F = $ MSA/MSE; RR: $F \geq 4.26$

$f = 3.6669/0.1880 = 19.51 \geq 4.26$

There is evidence to suggest at least two population mean amounts of saturated fat per serving are different.

b. df 4 and 44, $Q_{0.05} = 3.777$

Difference	Tukey CI	Significantly different
$\mu_1 - \mu_2$	$(\ 0.39,\ 1.34)$	Yes
$\mu_1 - \mu_3$	$(-0.18,\ 0.76)$	No
$\mu_1 - \mu_4$	$(\ 0.75,\ 1.70)$	Yes
$\mu_2 - \mu_3$	$(-1.05, -0.10)$	Yes
$\mu_2 - \mu_4$	$(-0.11,\ 0.83)$	No
$\mu_3 - \mu_4$	$(\ 0.46,\ 1.41)$	Yes

c. All differences are consistent, except between 68% fat and 40% fat. We would expect evidence to suggest these two population means are different. The Tukey CI just barely includes 0.

Section 11.3: Two-Way ANOVA

11.47 a. $a = 2$, $b = 3$, $n = 3$

b. $t_{11.} = 21$, $t_{12.} = 31$, $t_{13.} = 37$,
$t_{21.} = 28$, $t_{22.} = 32$, $t_{23.} = 48$

c. $t_{1..} = 89$, $t_{2..} = 108$,
$t_{.1.} = 49$, $t_{.2.} = 63$, $t_{.3.} = 85$, $t_{...} = 197$

11.49 a. SST $= 481.880$, SSA $= 160.820$, SSB $= 76.8267$, SS(AB) $= 45.48$, SSE $= 198.7533$

b. MSA $= 160.82/3 = 53.6067$,
MSB $= 76.8267/2 = 38.4133$,
MS(AB) $= 45.48/6 = 7.58$,
MSE $= 198.7533/24 = 8.2814$

c. $f_A = 53.6067/8.2814 = 6.47$,
$f_B = 38.4133/8.2814 = 4.64$,
$f_{AB} = 7.58/8.2814 = 0.92$

d. Interaction: df 6 and 24, $F_{0.05} = 2.51$
$f_{AB} = 0.92$; there is no evidence of interaction.

Factor A: df 3 and 24, $F_{0.05} = 3.01$
$f_A = 6.47 \geq 3.01$; there is evidence of an effect due to factor A.

Factor B: df 2 and 24, $F_{0.05} = 3.40$
$f_B = 4.64 \geq 3.40$; there is evidence of an effect due to factor B.

11.51

Source of variation	Sum of squares	Degrees of freedom	Mean square	F	p value
Factor A	162.64	4	40.66	2.53	0.0476
Factor B	156.54	2	78.27	4.86	0.0103
Interaction	144.23	8	18.03	1.12	0.3596
Error	1206.87	75	16.09		
Total	1670.28	89			

Interaction: df 8 and 75, $F_{0.05} = 2.06$
$f_{AB} = 1.12$; there is no evidence of interaction.

Factor A: df 4 and 75, $F_{0.05} = 2.49$
$f_A = 2.53$; there is evidence of an effect due to factor A.

Factor B: df 2 and 75, $F_{0.05} = 3.12$
$f_B = 4.86 \geq 3.12$; there is evidence of an effect due to factor B.

11.53 a.

Source of variation	Sum of squares	Degrees of freedom	Mean square	F	p value
Type	90.14	2	45.07	3.02	0.0580
Restaurant	266.49	3	88.83	5.96	0.0015
Interaction	56.59	6	9.43	0.63	0.7054
Error	715.60	48	14.91		
Total	1128.82	59			

b. $abn - 1 = 59$
$\Rightarrow$ total number of observations $= 60$

c. Interaction: df 6 and 48, $F_{0.05} = 2.29$
$f_{AB} = 0.63$; there is no evidence of interaction. The

other two hypothesis tests can be conducted as usual.

d. Factor A: df 2 and 48, $F_{0.05} = 3.19$

$f_A = 3.02$; there is no evidence of an effect due to type.

Factor B: df 3 and 48, $F_{0.05} = 2.80$

$f_B = 5.96$; there is evidence of an effect due to restaurant.

11.55

Source of variation	Sum of squares	Degrees of freedom	Mean square	F	p value
Megapixels	86.201	3	28.73	5.83	0.0039
Printer	2.101	1	2.10	0.43	0.5182
Interaction	6.011	3	2.00	0.41	0.7473
Error	118.315	24	4.93		
Total	212.629	31			

Interaction: df 3 and 24, $F_{0.05} = 3.01$

$f_{AB} = 0.41$; there is no evidence of interaction.

Factor A: df 3 and 24, $F_{0.05} = 3.01$

$f_A = 5.83 \geq 3.01$; there is evidence of an effect due to megapixels.

Factor B: df 1 and 24, $F_{0.05} = 4.26$

$f_B = 0.43$; there is no evidence of an effect due to printer.

11.57

Source of variation	Sum of squares	Degrees of freedom	Mean square	F	p value
Office	30.25	2	15.13	0.40	0.6761
Type	216.00	1	216.00	5.66	0.0286
Interaction	2.25	2	1.13	0.03	0.9705
Error	687.50	18	38.19		
Total	936.00	23			

Interaction: df 2 and 18, $F_{0.05} = 3.55$

$f_{AB} = 0.03$; there is no evidence of interaction.

Factor A: df 2 and 18, $F_{0.05} = 3.55$

$f_A = 0.40$; there is no evidence of an effect due to office.

Factor B: df 1 and 18, $F_{0.05} = 4.41$

$f_B = 5.66 \geq 4.41$; there is evidence of an effect due to type of development.

11.59 a.

Source of variation	Sum of squares	Degrees of freedom	Mean square	F	p value
Cover	1.5409	3	0.5137	0.87	0.4770
State	3.1184	3	1.0395	1.76	0.1953
Interaction	7.1078	9	0.7898	1.34	0.2919
Error	9.4550	16	0.5909		
Total	21.2222	31			

Interaction: df 9 and 16, $F_{0.01} = 3.78$

$f_{AB} = 1.34$; there is no evidence of interaction.

b. Factor A: df 3 and 16, $F_{0.01} = 5.29$

$f_A = 0.87$; there is no evidence of an effect due to cover type.

Factor B: df 3 and 16, $F_{0.01} = 5.29$

$f_B = 1.76$; there is no evidence of an effect due to state.

11.61 $SSA = 465,241.39 - 465,038.44 = 202.95$
$SSB = 465,868.93 - 465,038.44 = 830.49$
$SS(AB) = 466,213.5 - 465,241.39 - 465,868.93$
$\qquad + 465,038.44 = 141.62$
$SSE = 1680.66 - 202.95 - 830.49 - 141.62 = 505.60$

Source of variation	Sum of squares	Degrees of freedom	Mean square	F	p value
Region	202.95	2	101.475	7.23	0.0023
Quality	830.49	3	276.83	19.71	0.0000
Interaction	141.62	6	23.6033	1.68	0.1541
Error	505.60	36	14.0444		
Total	1680.66	47			

Interaction: df 6 and 36, $F_{0.05} = 2.36$

$f_{AB} = 1.68$; there is no evidence of interaction.

Factor A: df 2 and 36, $F_{0.05} = 3.26$

$f_A = 7.23 \geq 3.26$; there is evidence of an effect due to region.

Factor B: df 3 and 36, $F_{0.05} = 2.87$

$f_B = 19.71 \geq 2.87$; there is evidence of an effect due to road marking quality.

11.63

Source of variation	Sum of squares	Degrees of freedom	Mean square	F	p value
Injury	68.89	4	17.2221	4.32	0.0034
State	4.76	2	2.3788	0.60	0.5514
Interaction	93.55	8	11.6936	2.94	0.0065
Error	298.72	75	3.9829		
Total	465.91	89			

Interaction, df 8 and 75, $F_{0.05} = 2.06$

$f_{AB} = 2.94 \geq 2.06$; there is no evidence of interaction.

Factor A: df 4 and 75, $F_{0.05} = 2.49$
$f_A = 4.32 \geq 2.49$; there is evidence of an effect due to injury type.

Factor B: df 2 and 75, $F_{0.05} = 3.12$
$f_B = 0.60$; the effect due to state is inconclusive.

Chapter Exercises

11.65 a.

Source of variation	Sum of squares	Degrees of freedom	Mean square	F	p value
Factor	32.295	4	8.0737	0.77	0.5489
Error	470.355	45	10.4523		
Total	502.650	49			

b. Five regions in Florida were considered. A total of 50 observations were obtained.

c. $f = 0.77$, $p = 0.5489$
There is no evidence to suggest at least two of the population mean boat-ramp angles are different.

11.67

Source of variation	Sum of squares	Degrees of freedom	Mean square	F	p value
Factor	3.7075	3	1.2358	1.67	0.1951
Error	20.6675	28	0.7381		
Total	24.3750	31			

H_0: $\mu_1 = \mu_2 = \mu_3 = \mu_4$;
H_a: $\mu_i \neq \mu_j$ for some $i \neq j$;
TS: $F = MSA/MSE$; RR: $F \geq 4.57$

$f = 1.67$, $p = 0.1951$
There is no evidence to suggest at least two of the population mean displacements are different.

11.69 df 5 and 120, $Q_{0.05} = 3.918$

Difference	Tukey CI	Significantly different
$\mu_1 - \mu_2$	$(-0.28, -0.02)$	Yes
$\mu_1 - \mu_3$	$(-0.19, 0.07)$	No
$\mu_1 - \mu_4$	$(-0.28, -0.02)$	Yes
$\mu_1 - \mu_5$	$(-0.29, -0.02)$	Yes
$\mu_2 - \mu_3$	$(-0.05, 0.21)$	No
$\mu_2 - \mu_4$	$(-0.13, 0.13)$	No
$\mu_2 - \mu_5$	$(-0.14, 0.12)$	No
$\mu_3 - \mu_4$	$(-0.21, 0.05)$	No
$\mu_3 - \mu_5$	$(-0.22, 0.04)$	No
$\mu_4 - \mu_5$	$(-0.14, 0.12)$	No

Graph of results:

$\overline{x}_1.$	$\overline{x}_3.$	$\overline{x}_2.$	$\overline{x}_4.$	$\overline{x}_5.$
0.5566	0.6190	0.7020	0.7023	0.7115

11.71

Source of variation	Sum of squares	Degrees of freedom	Mean square	F	p value
Group	0.0161	3	0.0054	0.15	0.9287
Tooth type	0.7200	1	0.7200	19.84	0.0002
Interaction	0.0827	3	0.0276	0.76	0.5276
Error	0.8709	24	0.0363		
Total	1.6896	31			

Interaction: df 3 and 24, $F_{0.05} = 3.01$
$f_{AB} = 0.76$; there is evidence of interaction.

Factor A: df 3 and 24, $F_{0.05} = 3.01$
$f_A = 0.15$; the effect due to group is inconclusive.

Factor B: df 1 and 24, $F_{0.05} = 4.26$
$f_B = 19.84 \geq 4.26$; there is evidence of an effect due to tooth type.

11.73 a. H_0: $\mu_1 = \mu_2 = \mu_3 = \mu_4 = \mu_5$;
H_a: $\mu_i \neq \mu_j$ for some $i \neq j$;
nl TS: $F = MSA/MSE$; RR: $F \geq 2.57$

$f = 505.9928/124.2791 = 4.07 \geq 2.57$

There is evidence to suggest at least two of the population mean catfish weights are different.

b. df 5 and 43, $Q_{0.01} = 4.913$

Difference	Tukey CI		Significantly different
$\mu_1 - \mu_2$	(−33.32,	2.63)	No
$\mu_1 - \mu_3$	(−33.24,	3.46)	No
$\mu_1 - \mu_4$	(−36.41,	−1.09)	Yes
$\mu_1 - \mu_5$	(−38.76,	1.28)	No
$\mu_2 - \mu_3$	(−16.45,	17.36)	No
$\mu_2 - \mu_4$	(−19.55,	12.74)	No
$\mu_2 - \mu_5$	(−22.10,	15.30)	No
$\mu_3 - \mu_4$	(−20.42,	12.70)	No
$\mu_3 - \mu_5$	(−22.91,	15.21)	No
$\mu_4 - \mu_5$	(−18.39,	18.40)	No

Graph of results:

$\overline{x}_{1.}$	$\overline{x}_{3.}$	$\overline{x}_{2.}$	$\overline{x}_{5.}$	$\overline{x}_{4.}$
20.70	35.59	36.05	39.44	39.45

Recommend Campground (okay, anywhere but Hill's Landing). On average, the largest catfish are caught at this location.

11.75

SSA = 6,216,941.75 − 6,207,572.25 = 9369.50

SSB = 62,29,279.42 − 6,207,572.25 = 21,707.17

SS(AB) = 6,257,223.75 − 6,216,941.75

$\qquad$ − 6,229,279.42 + 6,207,572.25

$\qquad$ = 18,574.83

SSE = 53500.88

a. Interaction: df 4 and 27, $F_{0.05} = 2.73$

$f_{AB} = 2.34$; there is no evidence of interaction.

b. Factor A: df 2 and 27, $F_{0.05} = 3.35$

$f_A = 2.36$; there is no evidence of an effect due to species ID.

c. Factor B: df 2 and 27, $F_{0.05} = 3.35$

$f_B = 5.48 \geq 3.35$; there is evidence of an effect due to age.

Chapter 12: Correlation and Linear Regression

Section 12.1: Simple Linear Regression

12.1 a. Appropriate; the slope is negative. **b.** Not appropriate; the relationship is not linear.
c. Appropriate; the slope is approximately zero.
d. Not appropriate; there is no obvious linear relationship.

12.3 a. $E(Y \mid 150) = 75.0 + 3.6(150) = 615$

b. Change in the dependent variable is 3.6, the coefficient on the independent variable.

c. For $x = 120$, $Y \sim N(507, 36)$
$P(Y < 500) = P(Z < -1.1667) = 0.1217$

12.5 a.
$\widehat{\beta}_1 = \frac{(12)(-246,677) - (460.53)(-6349.7)}{(12)(17,875.1) - (460.53)^2} = -14.8744$
$\widehat{\beta}_0 = \frac{-6349.7 - (-14.8744)(460.53)}{12} = 41.7044$
$y = 41.7004 - 14.8744x$

b. $E(Y \mid 41) = 41.7044 - 14.8744(41) = -568.15$

12.7 a. Scatter plot and regression line:

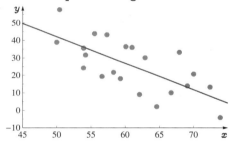

A simple linear regression model seems reasonable. The points appear to fall near a straight line.
b. $\widehat{\beta}_1 = \frac{(21)(31065.36) - (1280.1)(534.4)}{(21)(79026.77) - (1280.1)^2} = -1.5169$
$\widehat{\beta}_0 = \frac{534.4 - (-1.5169)(1280.1)}{21} = 117.9145$
$y = 117.9145 - 1.5169x$

Graph of the regression line added to scatter plot above.

c. $E(Y \mid 62) = 117.9145 - 1.5169(62) = 23.8667$

d.

Source of variation	Sum of squares	Degrees of freedom	Mean square	F
Regression	2290.75	1	2290.75	18.18
Error	2393.78	19	125.99	
Total	4684.53	20		

12.9 a. $y = 8.3 - 0.09(65) = 2.45$

b. $(-0.09)(10) = -0.90$

c. For $x = 80$, $Y \sim N(1.1, 0.25^2)$
$P(Y < 1) = P(Z < -0.4) = 0.3446$

12.11 a. $\widehat{\beta}_1 = \frac{(14)(19.61) - (8.6)(37.4)}{(14)(7.26) - (8.6)^2} = -1.7016$
$\widehat{\beta}_0 = \frac{37.4 - (-1.7016)(8.6)}{14} = 3.7167$
$y = 3.7167 - 1.7016x$

b. $E(Y \mid 1.25) = 3.7167 - 1.7016(1.25) = 1.5897$

12.13 a. $\widehat{\beta}_1 = \frac{(20)(545.653) - (50.95)(214.4)}{(20)(130.9151) - (50.95)^2} = -0.4741$
$\widehat{\beta}_0 = \frac{214.4 - (-0.4741)(50.95)}{20} = 11.9278$
$y = 11.9278 - 0.4741x$

b. $E(Y \mid 2.65) = 11.9278 - 0.4741(2.65) = 10.6714$

c. $y = 11.9279 - 0.4741(2.89) = 10.5577$

12.15 a. Scatter plot and estimated regression line:

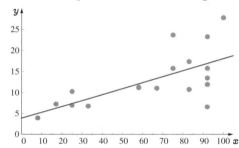

b. $\widehat{\beta}_1 = \frac{(18)(17494.4) - (1167.0)(235.6)}{(18)(91257.0) - (1167.0)^2} = 0.1423$
$\widehat{\beta}_0 = \frac{235.6 - (0.1423)(1167.0)}{18} = 3.8619$
$y = 3.8619 + 0.1423x$

c. $y = 3.8619 + 0.1423(75) = 14.5344$

12.17 a. $\widehat{\beta}_1 = \frac{(10)(5976.9) - (228.0)(259.1)}{(10)(5254.0) - (228.0)^2} = 1.2486$
$\widehat{\beta}_0 = \frac{259.1 - (1.2486)(228.0)}{10} = -2.5572$
$y = -2.5572 + 1.2486x$

b. ANOVA Summary Table:

Source of variation	Sum of squares	Degrees of freedom	Mean square	F
Regression	86.68	1	86.68	8.20
Error	84.55	8	10.57	
Total	171.23	9		

c. $y = -2.5572 + 1.2486(30) = 34.90$

12.19 a. $\widehat{\beta}_1 = \frac{(24)(384.0) - (219.0)(39.3)}{(24)(2093.0) - (219.0)^2} = 0.2683$
$\widehat{\beta}_0 = \frac{39.3 - (0.2683)(219.0)}{24} = -0.8107$
$y = -0.8107 + 0.2683x$

b. ANOVA Summary Table:

Source of variation	Sum of squares	Degrees of freedom	Mean square	F
Regression	6.81	1	6.81	14.95
Error	10.02	22	0.46	
Total	16.83	23		

c. $r^2 = 6.81/16.83 = 0.405$. Approximately 40% of the variation in the data is explained by the regression model.

d. Solve $-0.8107 + 0.2683x = 10$ for x
$\Rightarrow x = 40.2933$

12.21

a. $\widehat{\beta}_1 = \frac{(51)(67212.48) - (1139.6)(3699.7)}{(51)(77859.14) - (1139.6)^2} = -0.2950$

$\widehat{\beta}_0 = \frac{3699.7 - (-0.2950)(1139.6)}{51} = 79.1355$

$y = 79.1355 - 0.2950x$

b. ANOVA Summary Table:

Source of variation	Sum of squares	Degrees of freedom	Mean square	F
Regression	4560.38	1	4560.38	153.03
Error	1460.20	49	29.80	
Total	6020.59	50		

c. $r^2 = 4560.38/6020.59 = 0.7575$

d. $E(Y \mid 15.5) = 79.1355 - 0.2950(15.5) = 74.5626$

Section 12.2: Hypothesis Tests and Correlation

12.23 a. ANOVA Summary Table:

Source of variation	Sum of squares	Degrees of freedom	Mean square	F	p value
Regression	11691.9	1	11691.90	2.58	0.1219
Error	104372.1	23	4537.92		
Total	116064.0	24			

b. H_0: There is no significant linear relationship.
H_a: There is a significant linear relationship.
TS: $F = \text{MSR}/\text{MSE}$; RR: $F \geq 4.28$

$f = 11691.9/4537.92 = 2.58$

There is no evidence of a significant linear relationship.

c. $r^2 = 11691.9/116064.0 = 0.1007$

d. $r = \sqrt{0.1007} = 0.3174$ or -0.3174
There is no way to determine whether r is positive or negative from the ANOVA table.

12.25 a. H_0: $\beta_1 = 0$; H_a: $\beta_1 \neq 0$;
TS: $T = B_1/S_{B_1}$; RR: $|T| \geq 2.0796$
$t = \frac{4.4285}{\sqrt{561.088/151.086}} = 2.2980 \geq 2.0796$
There is evidence to suggest that $\beta_1 \neq 0$, the regression line is significant.

b. $\widehat{\beta}_1 \pm t_{0.025} \, s_{B_1}$

$= 4.4285 \pm (2.0796)\sqrt{\frac{561.088}{151.086}}$

$= (0.4209, 8.4361)$

c. There is evidence to suggest that $\beta_1 \neq 0$ since the CI does not include 0.

12.27 a. H_0: There is no significant linear relationship.
H_a: There is a significant linear relationship.
TS: $F = \text{MSR}/\text{MSE}$; RR: $F \geq 4.60$

$f = 1155.9/279.4871 = 4.14$

There is no evidence of a significant linear relationship.

df 1 and 14, $4.14 < 4.60$
$f < F_{0.05} \Rightarrow p > 0.05$
$p = P(F \geq 4.14) = 0.0614$

b. H_0: $\beta_1 = 0$; H_a: $\beta_1 \neq 0$;
TS: $T = B_1/S_{B_1}$; RR: $|T| \geq 2.1448$
$t = \frac{0.7640}{\sqrt{279.4871/1980.24}} = 2.0337$
There is no evidence to suggest that β_1 is different from 0.

$|t| = 2.0337$, df $= 14$
$1.7613 \leq 2.0337 \leq 2.1448$
$t_{0.05} \leq t \leq t_{0.025}$
$0.025 \leq p/2 \leq 0.05$
$0.05 \leq p \leq 0.10$
$p = 2P(T \geq 2.0337) = 0.0614$

c. $t^2 = (2.0337)^2 = 4.14 = f$

d. The p values are the same. These two hypothesis tests are testing the same null hypothesis.

12.29

a. $\widehat{\beta}_1 = \frac{(6)(-67105.38) - (237.0)(-1677.4)}{(6)(9457.88) - (237.0)^2} = -8.7993$

$\widehat{\beta}_0 = \frac{-1677.4 - (-8.7993)(237.0)}{6} = 68.0071$

$y = 68.0071 - 8.7993x$

ANOVA Summary Table:

Source of variation	Sum of squares	Degrees of freedom	Mean square	F	p value
Regression	7462.54	1	7462.54	10.35	0.0324
Error	2884.77	4	721.19		
Total	10347.31	5			

b. H_0: There is no significant linear relationship.
H_a: There is a significant linear relationship.
TS: $F = \text{MSR}/\text{MSE}$; RR: $F \geq 7.71$

$f = 7462.54/721.19 = 10.35 \geq 7.71$

There is evidence of a significant linear relationship.

c. $r^2 = 7462.54/10347.31 = 0.7212$

d. $r = \dfrac{-67105.38}{\sqrt{(9457.88)(479292.44)}} = -0.8492$

Since $r < 0$, the relationship between the two variables is negative. $\widehat{\beta}_1 < 0$ supports this answer, that the relationship is negative.

e. $r^2 = (-0.8492)^2 = 0.7212$

12.31 a. $\widehat{\beta}_1 = \dfrac{(15)(30.72)-(9.09)(50.44)}{(15)(5.73)-(9.09)^2} = 0.6925$

$\widehat{\beta}_0 = \dfrac{50.44-(0.6925)(9.09)}{15} = 2.9430$

$y = 2.9430 + 0.6925x$

ANOVA Summary Table:

Source of variation	Sum of squares	Degrees of freedom	Mean square	F	p value
Regression	0.1062	1	0.1062	0.40	0.5403
Error	3.4909	13	0.2685		
Total	3.5971	14			

b. H_0: There is no significant linear relationship.
H_a: There is a significant linear relationship.
TS: $F = \text{MSR}/\text{MSE}$; RR: $F \geq 9.07$

$f = 0.1062/0.2685 = 0.40$

There is no evidence of a significant linear relationship.

c. $\widehat{\beta}_0 \pm t_{0.025} s_{B_1}$

$= 2.9430 \pm (2.1604)\sqrt{\dfrac{(0.2685)(5.73)}{(15)(0.2215)}}$

$= (1.4728, 4.4132)$

There is evidence to suggest $\beta_1 \neq 0$ since the CI does not include 0.

12.33 a. $r = \dfrac{-2.8600}{\sqrt{(8.6786)(2.2771)}} = -0.6434$

b. There is a negative linear relationship between mean annual temperature and depth of permafrost layer. As the mean annual temperature increases, the depth of the permafrost layer decreases.

12.35 a. Scatter plot (child's income, y versus parent's income, x):

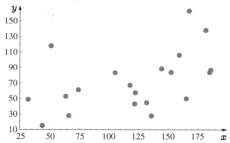

b. $r = \dfrac{17135.1850}{\sqrt{(46814.07)(27291.3495)}} = 0.4794$

c. The scatter plot and value of r support this theory. There is a slight positive linear relationship evident in the scatter plot and r is close to 0.5.

12.37 a. Scatter plot (coronary artery calcification, y versus anxiety measure, x):

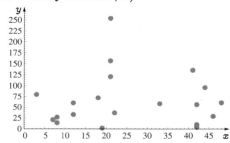

b. $r = \dfrac{180.5}{\sqrt{(4379.0)(73540.55)}} = 0.0101$

There is no clear relationship between these two variables.

12.39 a. $\widehat{\beta}_1 = \dfrac{(7)(1806400.0)-(1610.0)(7632.0)}{(7)(476750.0)-(1610)^2} = 0.4795$

$\widehat{\beta}_0 = \dfrac{7632.0-(0.4795)(1610.0)}{7} = 980.0067$

$y = 980.0067 + 0.4795x$

ANOVA Summary Table:

Source of variation	Sum of squares	Degrees of freedom	Mean square	F	p value
Regression	24472.35	1	24472.35	8.32	0.0344
Error	14705.08	5	2941.02		
Total	39177.43	6			

b. $r^2 = 24472.3/39177.4 = 0.6247$

c. $\widehat{\beta}_1 \pm t_{0.005} s_{B_1}$

$= 0.4795 \pm (4.0321)(0.1662)$

$= (-0.1907, 1.1497)$

d. H_0: $\beta_0 = 0$; H_a: $\beta_0 > 0$;
TS: B_0/S_{B_0}; RR: $T \geq 2.0150$ ($\alpha = 0.05$)

$t = 980.0067/43.3783 = 22.5921 \geq 2.0150$

There is evidence to suggest $\beta_0 > 0$. This suggests that even if the owner spends nothing on advertising in a week, he/she will still have a total weekly revenue greater than 0, close to 980.

12.41 a. $\widehat{\beta}_1 = \frac{(14)(32.3987) - (7.86)(48.17)}{(14)(6.3890) - (7.86)^2} = 2.7096$

$\widehat{\beta}_0 = \frac{48.17 - (2.7096)(7.86)}{14} = 1.9195$

$y = 1.9195 + 2.7096x$

b. $H_0: \beta_1 = 0$; $H_a: \beta_1 \neq 0$;
TS: $T = B_1/S_{B_1}$; RR: $|T| \geq 2.1788$

$t = 2.7096/1.1324 = 2.3928 \geq 2.1788$

There is evidence to suggest that β_1 is different from 0.

c. $E(Y \mid 0.55) = 1.9195 + 2.7096(0.55) = 3.4098$

d. $r^2 = 14.5092/44.9187 = 0.3230$

This model could be improved by obtaining more data.

12.43 a. $\widehat{\beta}_1 = \frac{(14)(48.518) - (64.4)(9.86)}{(14)(323.56) - (64.4)^2} = 0.1157$

$\widehat{\beta}_0 = \frac{9.86 - (0.1157)(64.4)}{14} = 0.1719$

$y = 0.1719 + 0.1157x$

ANOVA Summary Table:

Source of variation	Sum of squares	Degrees of freedom	Mean square	F	p value
Regression	0.3660	1	0.3660	11.67	0.0051
Error	0.3762	12	0.0313		
Total	0.7421	13			

b. $\widehat{\beta}_1 = \frac{(14)(702.9) - (994.0)(9.86)}{(14)(70902.0) - (994.0)^2} = 0.0087$

$\widehat{\beta}_0 = \frac{9.86 - (0.0087)(94.0)}{14} = 0.0895$

$y = 0.0895 + 0.0087x$

ANOVA Summary Table:

Source of variation	Sum of squares	Degrees of freedom	Mean square	F	p value
Regression	0.0246	1	0.0246	0.41	0.5334
Error	0.7176	12	0.0598		
Total	0.7421	13			

c. The evaporation rate and air velocity model is better because there is a significant linear relationship in this model. There is no significant linear relationship in the evaporation rate and relative humidity model.

Section 12.3: Inferences Concerning the Mean Value and an Observed Value of Y for $x = x^*$

12.45 a. $H_0: y^* = 20$; $H_a: y^* > 20$;
TS: $T = \frac{(B_0 + B_1 x^*) - y_0^*}{S\sqrt{(1/n) + [(x^* - \overline{x})^2/S_{xx}]}}$; RR: $T \geq 1.7459$

$t = \frac{20.376 - 20.0}{(10.1622)\sqrt{(1/18) + [(16.2 - 15.3670)^2/138.14]}} = 0.1503$

There is no evidence to suggest the mean value of Y for $x = 16.2$ is greater than 20.

b. $H_0: y^* = 5$; $H_a: y^* \neq 5$;
TS: $T = \frac{(B_0 + B_1 x^*) - y_0^*}{S\sqrt{(1/n) + [(x^* - \overline{x})^2/S_{xx}]}}$; RR: $|T| \geq 2.9208$

$t = \frac{4.49 - 5.0}{(10.1622)\sqrt{(1/18) + [(11.5 - 15.3670)^2/138.14]}} = -0.1240$

There is no evidence to suggest the mean value of Y for $x = 11.5$ is different from 5.

12.47 a. $(\widehat{\beta}_0 + \widehat{\beta}_1 x^*) \pm t_{0.005} s \sqrt{1 + \frac{1}{n} + \frac{(x^* - \overline{x})^2}{S_{xx}}}$

$= 49.9712 \pm (4.0321)(16.6433)\sqrt{1 + \frac{1}{7} + \frac{(19.25 - 19.1)^2}{15.131}}$

$= (-20.1474, 123.4294)$

width $= 123.4294 - (-20.1474) = 143.5768$

b. $(\widehat{\beta}_0 + \widehat{\beta}_1 x^*) \pm t_{0.005} s \sqrt{1 + \frac{1}{n} + \frac{(x^* - \overline{x})^2}{S_{xx}}}$

$= 51.6410 \pm (4.0321)(16.6433)\sqrt{1 + \frac{1}{7} + \frac{(18.1 - 19.1)^2}{15.131}}$

$= (-23.8157, 123.7581)$

width $= 123.7581 - (-23.8157) = 147.5738$

c. 18.1 is farther from the mean than 19.25.

12.49 a. $\widehat{\beta}_1 = \frac{(15)(62707.86) - (467.9)(2056.0)}{(15)(14765.37) - (467.9)^2} = -8.3856$

$\widehat{\beta}_0 = \frac{2056.0 - (-8.3856)(467.9)}{15} = 398.6420$

$y = 398.6420 - 8.3856x$

ANOVA Summary Table:

Source of variation	Sum of squares	Degrees of freedom	Mean square	F	p value
Regression	11954.82	1	11954.82	13.51	0.0028
Error	11502.28	13	884.79		
Total	23457.09	14			

H_0: There is no significant linear relationship.
H_a: There is a significant linear relationship.
TS: $F = MSR/MSE$; RR: $F \geq 9.07$

$f = 11954.82/884.79 = 13.51 \geq 9.07$

There is evidence of a significant linear relationship.

b. $s = \sqrt{884.79} = 29.7454$

c. $(\widehat{\beta}_0 + \widehat{\beta}_1 x^*) \pm t_{0.025} s \sqrt{1 + \frac{1}{n} + \frac{(x^* - \overline{x})^2}{S_{xx}}}$

$= 88.3742 \pm$

$$(2.1604)(29.7454)\sqrt{1 + \tfrac{1}{15} + \tfrac{(37-31.1933)^2}{170.0093}}$$

$= (16.0985, 160.6499)$

It is unlikely that an observed value of Y will be greater than 170 since this PI is completely below 170.

12.51

a. $E(Y \mid 130) = 0.2007 + (0.00446)(130) = 0.6467$

b. $(\widehat{\beta}_0 + \widehat{\beta}_1 x^*) \pm t_{0.025} s\sqrt{\tfrac{1}{n} + \tfrac{(x^* - \bar{x})^2}{S_{xx}}}$

$$= 0.7805 \pm (2.0930)(0.2477)\sqrt{\tfrac{1}{21} + \tfrac{(130-103.085)^2}{12335.8}}$$

$= (0.6115, 0.9495)$

c. $H_0: y^* = 0.06$; $H_a: y^* > 0.06$;

TS: $T = \dfrac{(B_0 + B_1 x^*) - y_0^*}{S\sqrt{(1/n) + [(x^* - \bar{x})^2/S_{xx}]}}$; RR: $T \geq 2.5395$

$t = \dfrac{0.5575 - 0.06}{(0.2477)(0.3014)} = 6.6636 \geq 2.5395$

There is evidence to suggest the mean value of Y for $x = 80$ is greater than 0.06.

12.53 a. $\sum x_i = (11)(792.636) = 8718.996$

$\sum y_i = (11)(5.26) = 57.86$

$\sum x_i^2 = 178661 + \tfrac{1}{11}(8718.996)^2 = 7{,}089{,}651.113$

$\sum x_i y_i = 380.955 + \tfrac{1}{11}(8718.996)(57.86) = 46242.874$

$\widehat{\beta}_1 = \dfrac{(11)(46242.874) - (8718.996)(57.86)}{(11)(7{,}089{,}651.113) - (8718.996)^2} = 0.0021$

$\widehat{\beta}_0 = 5.26 - (0.0021)(792.636) = 3.5955$

$y = 3.5955 + 0.0021x$

b. ANOVA Summary Table:

Source of variation	Sum of squares	Degrees of freedom	Mean square	F	p value
Regression	0.8000	1	0.8000	0.15	0.7089
Error	48.4655	9	5.3851		
Total	49.2655	10			

H_0: There is no significant linear relationship.
H_a: There is a significant linear relationship.
TS: $F = \text{MSR/MSE}$; RR: $F \geq 5.12$ ($\alpha = 0.05$)
$f = 0.80/5.3851 = 0.15$

There is no evidence of a significant linear relationship. Annual rainfall does not help to explain the variation in asthma prevalence.

c. $(\widehat{\beta}_0 + \widehat{\beta}_1 x^*) \pm t_{0.025} s\sqrt{1 + \tfrac{1}{n} + \tfrac{(x^* - \bar{x})^2}{S_{xx}}}$

$$5.6955 \pm (2.2622)(2.3206)\sqrt{1 + \tfrac{1}{11} + \tfrac{(1000-792.636)^2}{178661}}$$

$= (-0.3622, 11.7532)$

This PI includes some negative numbers, which is odd since we cannot have a negative percentage of adults treated for asthma.

12.55 a. The independent variable is the skid resistance. The dependent variable is the accident rate.

b. Scatter plot (accident rate, y versus skid resistance, x):

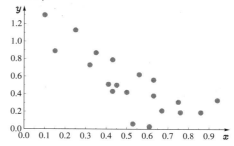

The relationship appears to be negative linear.

c. $\widehat{\beta}_1 = \dfrac{(20)(4.2335) - (10.33)(10.45)}{(20)(6.2829) - (10.33)^2} = -1.2285$

$\widehat{\beta}_0 = \dfrac{10.45 - (-1.2285)(10.33)}{20} = 1.1570$

$y = 1.1570 - 1.2285x$

d. $H_0: y^* = 0.60$; $H_a: y^* < 0.60$;

TS: $T = \dfrac{(B_0 + B_1 x^*) - y_0^*}{S\sqrt{(1/n) + [(x^* - \bar{x})^2/S_{xx}]}}$;

RR: $T \leq -1.7341$ ($\alpha = 0.05$)

$t = \dfrac{0.5428 - 0.6}{(0.2138)(0.2236)} = -1.1942$

There is no evidence to suggest the mean value of Y for $x = 0.50$ is less than 0.60.

12.57 a. Scatter plot (crimes, y versus temperature, x):

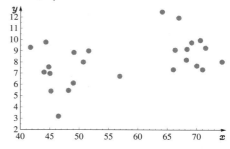

There is a weak positive linear relationship between average temperature and number of violent crimes.

b. $\widehat{\beta}_1 = \dfrac{(25)(12007.9040) - (1443.3)(203.91)}{(25)(86465.57) - (1443.3)^2} = 0.0751$

$\widehat{\beta}_0 = \dfrac{203.91 - (0.0751)(1443.3)}{25} = 3.8228$

$y = 3.8228 + 0.0751x$

c. $(\widehat{\beta}_0 + \widehat{\beta}_1 x^*) \pm t_{0.025} s\sqrt{1 + \tfrac{1}{n} + \tfrac{(x^* - \bar{x})^2}{S_{xx}}}$

$$= 9.8279 \pm (2.0687)(1.8657)\sqrt{1 + \tfrac{1}{25} + \tfrac{(80-57.732)^2}{3140.9744}}$$

$= (5.6039, 14.0519)$

d. $(\widehat{\beta}_0 + \widehat{\beta}_1 x^*) \pm t_{0.025} s\sqrt{1 + \tfrac{1}{n} + \tfrac{(x^* - \bar{x})^2}{S_{xx}}}$

$= 8.3266 \pm (2.0687)(1.8657)\sqrt{1 + \frac{1}{25} + \frac{(60-57.732)^2}{3140.9744}}$

$= (4.3877, 12.2656)$

12.59 a. ANOVA Summary Table:

Source of variation	Sum of squares	Degrees of freedom	Mean square	F	p value
Regression	853.50	1	853.50	8.99	0.0103
Error	1234.23	13	94.94		
Total	2087.73	14			

H_0: There is no significant linear relationship.
H_a: There is a significant linear relationship.
TS: $F = $ MSR/MSE; RR: $F \geq 4.67$ ($\alpha = 0.05$)

$f = 835.50/94.94 = 8.99 \geq 4.67$

There is evidence of a significant linear relationship. As CPI increases, so does ESI.

b. $E(Y \mid 6.7) = 26.432 + 4.546(6.7) = 56.8902$

c. $(\widehat{\beta}_0 + \widehat{\beta}_1 x^*) \pm t_{0.025} s \sqrt{1 + \frac{1}{n} + \frac{(x^* - \overline{x})^2}{S_{xx}}}$

$= 63.7092 \pm (2.1604)(35.1316)\sqrt{1 + \frac{1}{15} + \frac{(8.2-5.4333)^2}{41.2933}}$

$= (40.1554, 87.2630)$

Since the prediction interval is entirely below 90, it is unlikely a randomly selected country with a CIP of 8.2 will have an ESI greater than 90.

12.61 a. Scatter plot (length of stay, y versus number of patients per registered nurse, x):

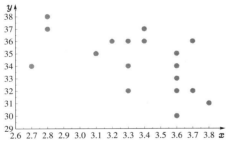

There is a weak negative linear relationship.

b. $\widehat{\beta}_1 = \frac{(21)(2419.8)-(70.3)(725.0)}{(21)(237.29)-(70.3)^2} = -3.70$

$\widehat{\beta}_0 = \frac{725.0-(-3.70)(70.3)}{21} = 46.91$

$y = 46.91 - 3.70x$

c. $(\widehat{\beta}_0 + \widehat{\beta}_1 x^*) \pm t_{0.025} s \sqrt{1 + \frac{1}{n} + \frac{(x^* - \overline{x})^2}{S_{xx}}}$

$= 33.22 \pm (2.8609)(1.8710)\sqrt{1 + \frac{1}{21} + \frac{(3.7-3.3476)^2}{1.9524}}$

$= (27.5775, 38.8625)$

d. $(\widehat{\beta}_0 + \widehat{\beta}_1 x^*) \pm t_{0.025} s \sqrt{1 + \frac{1}{n} + \frac{(x^* - \overline{x})^2}{S_{xx}}}$

$34.70 \pm (2.8609)(1.8710)\sqrt{1 + \frac{1}{21} + \frac{(3.3-3.3476)^2}{1.9524}}$

$= (29.2183, 40.1817)$

e. The prediction interval in part (c) is wider than the prediction interval in part (d) because 3.7 is farther from the mean than 3.3

Section 12.4: Regression Diagnostics

12.63

a.

x_i	y_i	$\widehat{y}_i$	$\widehat{e}_i$
1.44	8.59	9.5798	-0.9898
1.98	8.40	11.4005	-3.0005
1.69	13.60	10.4227	3.1773
1.29	9.25	9.0740	0.1760
1.16	11.16	8.6357	2.5243
1.66	11.03	10.3215	0.7085
1.32	9.12	9.1752	-0.0552
1.94	13.30	11.2656	2.0344
1.42	8.77	9.5123	-0.7423
1.66	11.34	10.3215	1.0185
1.89	10.76	11.0970	-0.3370
1.31	5.58	9.1415	-3.5615
1.40	7.43	9.4449	-2.0149
1.04	8.64	8.2311	0.4089
1.32	8.89	9.1752	-0.2852
1.31	10.08	9.1415	0.9386

b. $\sum \widehat{e}_i = -0.9898 + \cdots + 0.9386 = 0.0001$

12.65 a. Normal probability plot:

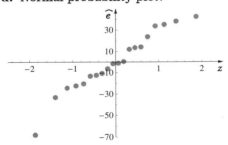

b. There is some evidence to suggest the random error terms are not normal. There appears to be an outlier and the points are slightly wavy.

12.67 a. There is evidence to suggest a violation in the regression model assumptions. The residual plot suggests the relationship between these variables in not linear.

b. There is evidence to suggest a violation in the regression model assumptions. The residual plot suggests the variance is not constant (and that the relationship may be nonlinear).

c. There is evidence to suggest a violation in the regression model assumptions. The residual plot suggests the relationship between these variables is not linear.

d. There is some evidence to suggest a violation in the regression model assumptions. The residual plot suggests the variance is not constant and increases as x increases.

12.69

a. $\widehat{\beta}_1 = \frac{(30)(439187.01)-(1981.0)(6901.1)}{(30)(135486.34)-(1981.0)^2} = -3.5333$

$\widehat{\beta}_0 = \frac{6901.1-(-3.5333)(1981.0)}{30} = 463.3508$

$y = 463.3508 - 3.5333x$

b.

x_i	y_i	$\widehat{y}_i$	$\widehat{e}_i$
71.5	51.7	210.7198	-159.0214
55.0	283.8	269.0193	14.7795
87.8	366.2	153.1271	213.0710
86.8	347.9	156.6604	191.2377
68.8	87.7	220.2598	-132.5613
68.4	57.5	221.6731	-164.1746
69.6	50.6	217.4331	-166.8346
82.1	203.7	173.2669	30.4313
89.4	424.4	147.4738	276.9243
71.1	79.0	212.1332	-133.1347
84.6	271.7	164.4336	107.2645
56.2	262.3	264.7793	-2.4806
69.6	70.1	217.4331	-147.3346
75.4	78.8	196.9400	-118.1416
57.4	212.6	260.5394	-47.9406
53.0	344.9	276.0859	68.8129
68.8	66.5	220.2598	-153.7613
51.7	403.4	280.6792	122.7197
51.5	394.9	281.3859	113.5130
57.4	216.0	260.5394	-44.5406
68.3	50.5	222.0264	-171.5279
50.7	427.6	284.2125	143.3864
81.6	184.4	175.0335	9.3647
51.0	414.8	283.1525	131.6464
54.9	284.2	269.3726	14.8262
53.4	341.3	274.6726	66.6262
51.8	397.4	280.3259	117.0730
76.3	86.6	193.7600	-107.1617
52.9	359.1	276.4392	82.6596
64.0	81.5	237.2196	-155.7210

c. Scatter plot of residuals versus x:

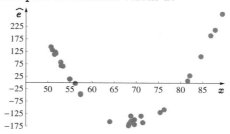

There is evidence to suggest a violation in the simple linear regression assumptions. There is a distinct curve in the residual plot.

12.71 a. Normal probability plot:

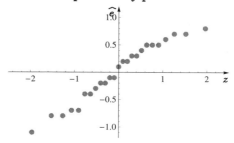

b. There is some evidence of non-normality. Each end of the plot flattens out slightly.

12.73 a. $y = 5.1677 + 0.000119x$

x_i	y_i	$\widehat{y}_i$	$\widehat{e}_i$
816932.0	133.0	102.5089	30.4911
1001121.0	140.0	124.4558	15.5442
1038437.0	139.0	128.9022	10.0978
965353.0	133.0	120.1939	12.8061
951545.0	137.0	118.5486	18.4514
928799.0	99.0	115.8383	-16.8383
991305.0	138.0	123.2862	14.7138
854258.0	92.0	106.9564	-14.9564
904858.0	89.0	112.9857	-23.9857
955466.0	133.0	119.0158	13.9842
833868.0	132.0	104.5269	27.4731
1079549.0	141.0	133.8009	7.1991
924059.0	135.0	115.2735	19.7265
856472.0	140.0	107.2202	32.7798
878897.0	96.0	109.8923	-13.8923
865363.0	83.0	108.2796	-25.2796
852244.0	132.0	106.7165	25.2835
945088.0	100.0	117.7792	-17.7792
808020.0	101.0	101.4470	-0.4470
889083.0	80.0	111.1060	-31.1060
892420.0	83.0	111.5036	-28.5036
905940.0	97.0	113.1146	-16.1146
790619.0	135.0	99.3736	35.6264
955003.0	139.0	118.9607	20.0393
831772.0	91.0	104.2771	-13.2771
935494.0	141.0	116.6361	24.3639
798612.0	85.0	100.3260	-15.3260
1062462.0	103.0	131.7649	-28.7649
793549.0	77.0	99.7227	-22.7227
866662.0	130.0	108.4344	21.5656
857782.0	133.0	107.3763	25.6237
949589.0	144.0	118.3156	25.6844
997925.0	103.0	124.0750	-21.0750
879987.0	90.0	110.0222	-20.0222
834344.0	83.0	104.5836	-21.5836
948066.0	133.0	118.1341	14.8659
949395.0	140.0	118.2924	21.7076
893983.0	88.0	111.6898	-23.6898
930016.0	81.0	115.9833	-34.9833
935863.0	89.0	116.6800	-27.6800

b. Normal probability plot:

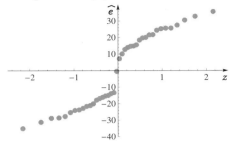

Scatter plot of residuals versus x:

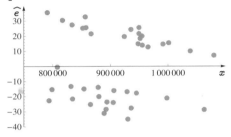

c. There is evidence to suggest the simple linear regression assumptions are invalid. There is a distinct nonlinear pattern in the normal probability plot and in the plot of the residuals versus the predictor variable.

12.75

a.

x_i	y_i	$\widehat{y}_i$	$\widehat{e}_i$
415.6	18.8	19.3180	-0.5180
362.2	20.3	18.5256	1.7744
529.9	20.0	21.0140	-1.0140
390.3	21.2	18.9425	2.2575
326.2	18.9	17.9914	0.9086
164.5	14.7	15.5920	-0.8920
423.8	20.1	19.4396	0.6604
169.3	14.3	15.6632	-1.3632
113.7	10.0	14.8381	-4.8381
275.0	19.2	17.2316	1.9684
265.1	18.1	17.0847	1.0153
597.6	18.8	22.0186	-3.2186
194.6	16.8	16.0386	0.7614
519.1	20.5	20.8538	-0.3538
585.7	21.7	21.8420	-0.1420
213.9	18.9	16.3250	2.5750
231.8	16.9	16.5906	0.3094
143.6	13.4	15.2818	-1.8818
471.1	20.2	20.1415	0.0585
223.5	18.4	16.4674	1.9326

$$\sum \widehat{e}_i = -0.5180 + \cdots + 1.9326 = 0.0000$$

b. Scatter plot of residuals versus x:

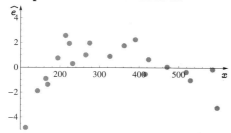

There is evidence to suggest the simple linear regression assumptions are violated. There is a pattern in this graph.

12.77

a. $\widehat{\beta}_1 = \frac{(10)(335992.0)-(283.9)(11505.0)}{(10)(8704.53)-(283.9)^2} = 14.5283$

$\widehat{\beta}_0 = \frac{11505.0-(14.5283)(283.9)}{10} = 738.0426$

$y = 738.0426 + 14.5283x$

x_i	y_i	$\widehat{y}_i$	$\widehat{e}_i$
29.8	1127	1170.9849	-43.9849
26.3	1105	1120.1359	-15.1359
26.0	1187	1115.7774	71.2226
37.2	1334	1278.4940	55.5060
30.1	1067	1175.3433	-108.3433
20.9	1094	1041.6833	52.3167
27.5	1115	1137.5698	-22.5698
39.8	1228	1316.2675	-88.2675
10.7	871	893.4950	-22.4950
35.6	1377	1255.2488	121.7512

b. Normal probability plot:

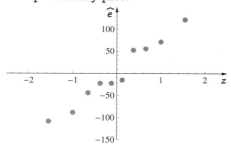

Scatter plot of residuals versus x:

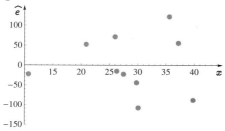

c. There is no overwhelming evidence that the simple linear regression assumptions are invalid.

There is a possible outlier, but the number of observations is small.

12.79 a. $\widehat{\beta}_1 = \frac{(20)(262799.0)-(1288.0)(4060.0)}{(20)(88418.0)-(1288.0)^2} = 0.2440$

$\widehat{\beta}_0 = \frac{4060.0-(0.2440)(1288.0)}{20} = 187.2849$

$y = 187.2849 + 0.2440x$

x_i	y_i	$\widehat{y}_i$	$\widehat{e}_i$
58	200	201.4383	-1.4383
73	201	205.0986	-4.0986
56	201	200.9502	0.0498
74	203	205.3426	-2.3426
43	202	197.7779	4.2221
46	199	198.5100	0.4900
64	199	202.9024	-3.9024
95	211	210.4671	0.5329
44	200	198.0219	1.9781
78	203	206.3187	-3.3187
77	201	206.0747	-5.0747
52	200	199.9741	0.0259
60	200	201.9263	-1.9263
98	226	211.1992	14.8008
46	202	198.5100	3.4900
86	205	208.2709	-3.2709
56	202	200.9502	1.0498
61	201	202.1703	-1.1703
76	203	205.8307	-2.8307
45	201	198.2660	2.7340

b. H_0: There is no significant linear relationship.
H_a: There is a significant linear relationship.
TS: $F = \text{MSR}/\text{MSE}$; RR: $F \geq 8.29$

$f = 325.77/20.13 = 16.19 \geq 8.29$

There is evidence of a significant linear relationship.

c. Normal probability plot:

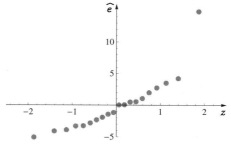

Scatter plot of residuals versus x:

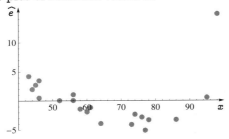

d. There is evidence to suggest the simple linear regression assumptions are invalid. Both plots suggest there is an outlier, and the residuals versus the predictor variable plot suggests a parabolic pattern. We might try excluding the outlier from the data set, or adding a quadratic term to the model.

12.81

a. $\widehat{\beta}_1 = \dfrac{(30)(9449.48)-(89.24)(2791.8)}{(30)(311.6430)-(89.24)^2} = 24.7881$

$\widehat{\beta}_0 = \dfrac{2791.8-(24.7881)(89.24)}{30} = 19.3238$

$y = 19.3238 + 24.7881x$

x_i	y_i	$\widehat{y}_i$	$\widehat{e}_i$
3.67	128.6	110.2960	18.3040
1.83	106.2	64.6859	41.5141
1.12	159.8	47.0864	112.7136
4.96	109.7	142.2726	-32.5726
4.37	108.8	127.6476	-18.8476
1.95	35.2	67.6605	-32.4605
4.49	93.3	130.6222	-37.3222
4.55	145.6	132.1095	13.4905
4.91	153.1	141.0332	12.0668
4.76	189.8	137.3150	52.4850
3.08	97.4	95.6710	1.7290
3.36	109.8	102.6117	7.1883
3.81	143.5	113.7663	29.7337
2.94	42.8	92.2007	-49.4007
1.42	41.8	54.5228	-12.7228
3.25	82.9	99.8850	-16.9850
3.95	86.5	117.2366	-30.7366
1.57	57.9	58.2410	-0.3410
1.55	3.7	57.7453	-54.0453
2.22	125.6	74.3533	51.2467
2.87	110.8	90.4655	20.3345
2.82	88.9	89.2261	-0.3261
3.25	56.9	99.8850	-42.9850
1.89	41.9	66.1732	-24.2732
1.28	93.3	51.0525	42.2475
3.66	47.6	110.0481	-62.4481
2.56	88.9	82.7812	6.1188
1.31	2.0	51.7961	-49.7961
1.21	21.5	49.3173	-27.8173
4.63	218.0	134.0925	83.9075

b. H_0: There is no significant linear relationship.
H_a: There is a significant linear relationship.
TS: $F = \text{MSR}/\text{MSE}$; RR: $F \geq 13.50$

$f = 28377.51/1817.71 = 15.61 \geq 13.50$

There is evidence of a significant linear relationship.

c. Normal probability plot:

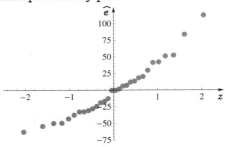

Scatter plot of residuals versus x:

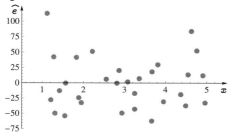

d. The graphs do not provide any evidence that the simple linear regression assumptions are invalid. The normal probability plot is approximately linear, and the residuals versus predictor variable plot exhibits no discernible pattern.

12.83

$$\sum_{i=1}^{n}(y_i - \widehat{y}_i) = \sum_{i=1}^{n}(y_i - (\widehat{\beta}_0 + \widehat{\beta}_1 x_i))$$

$$= \sum_{i=1}^{n}(y_i - (\overline{y} - \widehat{\beta}_1 \overline{x} + \widehat{\beta}_1 x_i))$$

$$= \sum_{i=1}^{n}(y_i - \overline{y}) - \widehat{\beta}_1 \sum_{i=1}^{n}(x_i - \overline{x})$$

$$= n\overline{y} - n\overline{y} - \widehat{\beta}_1(n\overline{x} - n\overline{x}) = 0$$

Section 12.5: Multiple Linear Regression

12.85 a. $E(Y) =$
$-10.7 + 5(1) - 14(2) - 23(2) + 6.7(4.5) = -49.55$

b. $(-14)(2) + (6.7)(-5) = -61.5$

c. $E(Y) =$
$-10.7 + 5(10) - 14(12.5) - 23(15) + 6.7(7) = -433.8$
$Y \sim N(-433.8, 24^2)$

$P(-460 \leq Y \leq 400)$
$= P(-1.0917 \leq Z \leq 34.7417)$
$= 1.0000 - 0.1375 = 0.8625$

12.87 a. Scatter plot of y versus x_1:

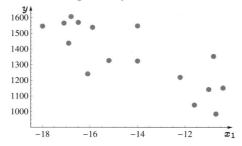

There is a slight negative linear relationship.

Scatter plot of y versus x_2:

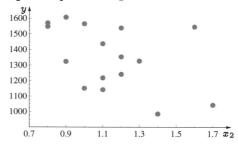

There is a very slight indication of a negative linear relationship.

Scatter plot of y versus x_3:

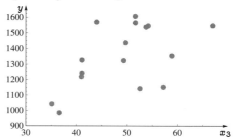

There is a slight indication of a positive linear relationship.

b.
$y = 221.9231 - 56.3497x_1 - 124.2215x_2 + 9.5798x_3$.
The sign of each estimated regression coefficient reflects the relationship in each scatter plot.

c.
$E(Y) = 221.9231 - 56.3497(-10) - 124.2215(1.35)$
$\qquad + 9.5798(52.6)$
$\qquad = 1121.6179$

12.89 a. $H_0: \beta_1 = \cdots = \beta_4 = 0$;
$H_a: \beta_i \neq 0$ for at least one i;
TS: $F = MSR/MSE$; RR: $F \geq 2.73$ ($\alpha = 0.05$)
$f = 11691/1738 = 6.73 \geq 2.73$

There is evidence to suggest that at least one of the regression coefficients is different from 0. The overall regression is significant.

b. β_2, β_3, and β_4 are significantly different from 0. In each hypothesis test to determine if the coefficient is significantly different from 0, $p \leq 0.05$. Therefore, x_2, x_3, and x_4 are significant predictor variables.

c. For df $= 27$, the critical value in each test is
$t_{0.05/8} = t_{0.0063} = 2.6763$

Using the Minitab output, β_3 is significantly different from 0, and therefore, x_3 is a significant predictor variable. This result is different from part (b).

12.91

a. $y = 114.4895 + 6.4722x_1 - 12.8017x_2$
$\qquad + 4.6091x_3 + 0.6409x_4$

$r^2 = 13305.955/17078.8733 = 0.7791$

b. β_1: $t = 4.3402$, $p = 0.0015$
β_2: $t = -0.5592$, $p = 0.5883$
β_3: $t = 1.4814$, $p = 0.1693$
β_4: $t = 2.9202$, $p = 0.0153$

Using $\alpha = 0.05$, x_1 and x_4 are significant predictors.

c. $Y_i = \beta_0 + \beta_1 x_{1i} + \beta_4 x_{4i} + E_i$

$y = 105.4656 + 5.1074x_1 + 0.6863x_4$

$r^2 = 12477.5792/17078.8733 = 0.7306$

d. The second model is better because it has fewer variables and the coefficient of determination is only slightly smaller.

12.93 a. df $= 20$, $t_{0.025} = 2.0860$

$y^* = -1.702 + 1.0138(1.5) - 1.0846(4.6)$
$\qquad + 3.4354(4.7) - 1.8904(4.9) + 1.3154(4.7)$
$\qquad = 7.8953$

$y^* \pm t_{0.025} s_{Y^*}$

$= 7.8953 \pm (2.0860)(1.049)$

$= (5.7072, 10.0835)$

We are 95% confident the true mean value of Y when $x = x^*$ lies in this interval.

b. $y^* \pm t_{0.025}\sqrt{s^2 + s_{Y^*}^2}$

$= 7.8953 \pm (2.0860)\sqrt{2.38509^2 + 1.049^2}$

$= (2.4602, 13.3305)$

We are 95% confident an observed value of Y when $x = x^*$ lies in this interval.

12.95

a. $y = 137.4024 + 0.0282x_1 - 4.4853x_2$

b. H_0: $\beta_1 = \beta_2 = 0$;
H_a: $\beta_i \neq 0$ for at least one i;
TS: $F = \text{MSR/MSE}$; RR: $F \geq 4.26$
$f = 144747.19/6747.59 = 21.45 \geq 4.26$

There is evidence to suggest that at least one of the regression coefficients is different from 0. The overall regression is significant.

c. $y = 137.4024 + 0.0282(10,000) - 4.4853(25)$
$\qquad = 307.2159$

12.97

a. $y = -3.1136 + 0.0554x_1 + 0.5777x_2 + 0.0028x_3$

b. H_0: $\beta_1 = \beta_2 = \beta_3 = 0$;
H_a: $\beta_i \neq 0$ for at least one i;
TS: $F = \text{MSR/MSE}$; RR: $F \geq 3.49$ $(\alpha = 0.05)$

$f = 0.6564/0.1670 = 3.93 \geq 3.49$

There is evidence to suggest that at least one of the regression coefficients is different from 0. The overall regression is significant.

β_1: $t = 0.0554/0.0224 = 2.4743$
$p = 0.0293$

β_2: $t = 0.5777/0.2017 = 2.8634$
$p = 0.0143$

β_3: $t = 0.0028/0.0064 = 0.4409$
$p = 0.6671$

The temperature of the solutions and the concentration of the solutions are the most important (significant) variables.

c. Normal probability plot:

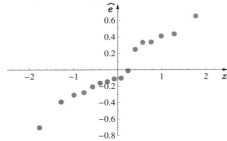

There is some evidence to suggest a violation in the multiple linear regression assumptions. The points in this plot are slightly nonlinear.

12.99

a. $y = 12297.6651 + 0.2596x_1 + 51.1428x_2$
$\qquad - 22.6309x_3 - 17.8406x_4$

b. H_0: $\beta_1 = \cdots = \beta_4 = 0$;
H_a: $\beta_i \neq 0$ for at least one i;
TS: $F = \text{MSR/MSE}$; RR: $F \geq 3.26$

$f = 15.17 \geq 3.26$

There is evidence to suggest that at least one of the regression coefficients is different from 0. The overall regression is significant.

$r^2 = 0.8465$

c. β_1: $t = 0.2596/0.1588 = 1.6341$
$p = 0.1305$

β_2: $t = 51.1428/24.2382 = 2.1100$
$p = 0.0586$

β_3: $t = -22.6309/20.6431 = -1.0963$
$p = 0.2964$

β_4: $t = -17.8406/24.4390 = -0.7300$

$p = 0.4806$

The only predictor variable that is close to significant is x_2.

d. $H_0: \beta_1 = 0.20$; $H_a: \beta_1 > 0.20$;

TS: $T = (B_1 - 0.20)/S_{B_1}$; RR: $T \geq 1.7823$

$t = \frac{0.2596 - 0.20}{0.1588} = 0.3753$

There is no evidence to suggest that $\beta_1 > 0.20$.

12.101 a. ANOVA Summary Table:

Source of variation	Sum of squares	Degrees of freedom	Mean square	F	p value
Regression	531.54	3	177.18	6.21	0.0035
Error	599.38	21	28.5419		
Total	1130.92	24			

$H_0: \beta_1 = \beta_2 = \beta_3 = 0$;
$H_a: \beta_i \neq 0$ for at least one i;
TS: $F = MSR/MSE$; RR: $F \geq 3.07$ ($\alpha = 0.05$)

$f = 177.18/28.5419 = 6.21 \geq 3.07$

There is evidence to suggest that at least one of the regression coefficients is different from 0. The overall regression is significant.

b. All three regression coefficients, β_1, β_2, and β_3, are significantly different from 0. In each hypothesis, test to determine if the coefficient is significantly different from 0, $p \leq 0.05$. Therefore, all three variables contribute to the overall significant regression.

c. df $= 21$, $t_{0.025} = 2.0796$

$y^* \pm t_{0.025} s_{Y^*}$

$= 55.63 \pm (2.0796)(1.44)$

$= (52.64, 58.62)$

$y^* \pm t_{0.025} s_{Y^*}$

$= 62.37 \pm (2.0796)(2.96)$

$= (56.21, 68.54)$

d. df $= 21$, $t_{0.025} = 2.0796$

$y^* \pm t_{0.025} \sqrt{s^2 + s_{Y^*}^2}$

$= 55.63 \pm (2.0796)\sqrt{5.34244^2 + 1.44^2}$

$= (44.12, 67.14)$

$y^* \pm t_{0.025} \sqrt{s^2 + s_{Y^*}^2}$

$= 62.37 \pm (2.0796)\sqrt{5.34244^2 + 2.96^2}$

$= (49.67, 75.08)$

e. x_1^* is closer to the *mean* than x_2^*.

12.103

a. $y = 197.1839 - 3.5821x_1 - 6.2638x_2$

b. $H_0: \beta_1 = \beta_2 = 0$;
$H_a: \beta_i \neq 0$ for at least one i;
TS: $F = MSR/MSE$; RR: $F \geq 4.74$

$f = 14554.5686/2000.5375 = 7.28 \geq 4.74$

There is evidence to suggest that at least one of the regression coefficients is different from 0. The overall regression is significant.

c. $r^2 = 29109.1372/43112.9 = 0.6752$

Approximately 68% of the variation in y is explained by this regression model.

d. β_1: $t = -3.5821/1.15109 = -2.3709$
$p = 0.0495 \leq 0.05$

β_2: $t = -6.2638/2.0925 = -2.9934$
$p = 0.0201 \leq 0.05$

Both regression coefficients are significantly different from 0. Therefore, x_1 and x_2 are both significant predictor variables.

e. Yes. The claim made by Hormel is believable. The overall regression is significant, and both variables contribute to the overall significance.

12.105 a. Scatter plot of y versus x:

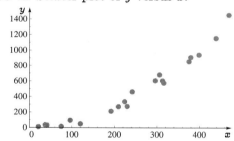

b. $\widehat{\beta}_1 = \frac{(20)(3357775.0) - (4779.0)(9577.0)}{(920)(1510143.0) - (4779.0)^2} = 2.9043$

$\widehat{\beta}_0 = \frac{9577.0 - (2.9043)(4779.0)}{20} = -215.1226$

$y = -215.1226 + 2.9043x$

ANOVA Summary Table:

Source of variation	Sum of squares	Degrees of freedom	Mean square	F	p value
Regression	3105671.6	1	3105671.6	201.69	< 0.0001
Error	277162.9	18	15397.9		
Total	3382834.5	19			

c. Small depths (≤ 150):

$y = 7.4516 + 0.4828x$

ANOVA Summary Table:

Source of variation	Sum of squares	Degrees of freedom	Mean square	F	p value
Regression	1720.68	1	1720.68	2.47	0.1915
Error	2792.15	4	698.04		
Total	4512.83	5			

Large depths (> 150):

$y = -598.5726 + 4.0383x$

ANOVA Summary Table:

Source of variation	Sum of squares	Degrees of freedom	Mean square	F	p value
Regression	1654572.7	1	1654572.7	309.29	< 0.0001
Error	64194.7	12	5349.6		
Total	1718767.4	13			

d. $y = 7.3195 + 0.0062x^2$

e. The model using x^2 seems to be the best. A model with x^2 appears to fit the scatter plot, and the value of r^2 for this model is 0.987, which is very high.

Chapter Exercises

12.107 a. $\widehat{\beta}_1 = \frac{(22)(3644.2)-(155.2)(392.0)}{(22)(1508.8)-(155.2)^2} = 2.1231$

$\widehat{\beta}_0 = \frac{392.0-(2.1231)(155.2)}{22} = 2.8408$

$y = 2.8408 + 2.1231x$

b. $E(Y \mid 10) = 2.8408 + 2.1231(10) = 24.0718$

c. $y = 2.8408 + 2.1231(2) = 7.0870$

12.109 a. $\widehat{\beta}_1 = \frac{(18)(27254.1)-(7777.0)(58.8)}{(18)(3674051.0)-(7777.0)^2} = 0.0059$

$\widehat{\beta}_0 = \frac{58.8-(0.0059)(7777.0)}{18} = 0.7218$

$y = 0.7218 + 0.0059x$

ANOVA Summary Table:

Source of variation	Sum of squares	Degrees of freedom	Mean square	F	p value
Regression	10.8922	1	10.8922	5.79	0.0285
Error	30.0878	16	1.8805		
Total	40.9800	17			

b. $r^2 = 10.8922/40.9800 = 0.2658$

c. $r = \frac{1849.2333}{\sqrt{(313954.9444)(40.9800)}} = 0.5156$

$r^2 = (0.5156)^2 = 0.2658$

d. Seagrass density is not a very good measure of water quality. The correlation is only moderate, the regression is barely significant, and the r^2 value is low.

12.111 a. Scatter plot (percentage of defoliation, y versus concentration of phosphorus, x) and estimated regression line:

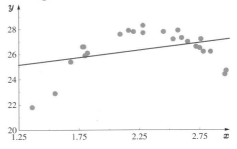

The relationship between x and y appears to be quadratic.

b. $\widehat{\beta}_1 = \frac{(24)(1467.972)-(55.37)(633.6)}{(24)(133.0157)-(55.37)^2} = 1.1767$

$\widehat{\beta}_0 = \frac{633.6-(1.1767)(55.37)}{24} = 23.6853$

$y = 23.6853 + 1.1767x$

H_0: There is no significant linear relationship.

H_a: There is a significant linear relationship.

TS: $F = MSR/MSE$; RR: $F \geq 4.30$

$f = 7.3001/2.3709 = 3.08$

There is no evidence of a significant linear relationship.

c.

x_i	y_i	$\widehat{y}_i$	$\widehat{e}_i$
2.76	27.2	26.9329	0.2671
2.78	26.2	26.9565	-0.7565
2.57	27.9	26.7094	1.1906
2.72	26.6	26.8859	-0.2859
2.84	26.2	27.0271	-0.8271
2.60	27.3	26.7447	0.5553
2.65	27.0	26.8035	0.1965
2.16	27.9	26.2269	1.6731
2.28	28.3	26.3681	1.9319
2.28	27.7	26.3681	1.3319
1.55	22.9	25.5092	-2.6092
2.20	27.8	26.2740	1.5260
1.82	26.1	25.8269	0.2731
2.96	24.4	27.1683	-2.7683
2.97	24.7	27.1800	-2.4800
1.68	25.4	25.6621	-0.2621
1.36	21.8	25.2856	-3.4856
1.78	26.6	25.7798	0.8202
2.45	27.8	26.5682	1.2318
1.80	25.9	25.8033	0.0967
2.75	26.5	26.9212	-0.4212
1.79	26.6	25.7916	0.8084
2.53	27.2	26.6623	0.5377
2.09	27.6	26.1446	1.4554

Scatter plot of residuals versus x:

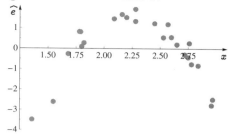

d. The graphs in part (a) and (c) suggest the model could be imporved by adding a quadratic term: x^2.

12.113 a. $\widehat{\beta}_1 = 111.877/185.725 = 0.6024$

$\widehat{\beta}_0 = 11.7654 - (0.6024)(5.6084) = 8.3870$

$y = 8.3870 + 0.6024x$

b. ANOVA Summary Table:

Source of variation	Sum of squares	Degrees of freedom	Mean square	F	p value
Regression	67.39	1	67.39	2.29	0.1370
Error	1414.44	48	29.47		
Total	1481.83	49			

c. H_0: $\beta_1 = 0$; H_a: $\beta_1 \neq 0$;

TS: $T = B_1/S_{B_1}$; RR: $|T| \geq 2.0106$

$t = \dfrac{0.6024}{\sqrt{29.47/185.725}} = 1.5123$

There is no evidence to suggest that $\beta_1 \neq 0$, the regression line is not significant.

d. The saying does not seem appropriate. There is no significant relationship between rating and price.

12.115 a. $\widehat{\beta}_1 = \dfrac{(10)(21.5190)-(12.62)(14.6)}{(10)(19.3416)-(12.62)^2} = 0.9059$

$\widehat{\beta}_0 = \dfrac{14.6-(0.9059)(12.62)}{10} = 0.3168$

$y = 0.3168 + 0.9059x$

ANOVA Summary Table:

Source of variation	Sum of squares	Degrees of freedom	Mean square	F	p value
Regression	2.8027	1	2.8027	17.23	0.0032
Error	1.3013	8	0.1627		
Total	4.1040	9			

b. df $= 8$, $t_{0.025} = 2.3060$

$(\widehat{\beta}_0 + \widehat{\beta}_1 x^*) \pm t_{0.005} s\sqrt{1 + \dfrac{1}{n} + \dfrac{(x^*-\overline{x})^2}{S_{xx}}}$

$= 1.7662 \pm (2.3060)(0.4033)\sqrt{1 + \dfrac{1}{10} + \dfrac{(1.6-1.262)^2}{3.4152}}$

$= (0.7760, 2.7564)$

c. H_0: $y^* = 1$; H_a: $y^* > 1$;

TS: $T = \dfrac{(B_0 + B_1 x^*) - y_0^*}{S\sqrt{(1/n)+[(x^*-\overline{x})^2/S_{xx}]}}$; RR: $T \geq 2.8965$

$t = \dfrac{1.0415-1}{(0.4033)\sqrt{\frac{1}{10}+\frac{(0.8-1.262)^2}{3.4152}}} = 0.2554$

There is no evidence to suggest the mean value of Y for $x = 0.8$ is greater than 1.

12.117 a. Scatter plot (cicada density, y versus mole density, x):

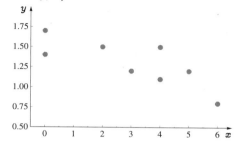

There appears to be a negative linear relationship.

b. $\widehat{\beta}_1 = \dfrac{(9)(33.8)-(28.0)(11.9)}{(9)(122.0)-(28.0)^2} = -0.0924$

$\widehat{\beta}_0 = \dfrac{11.9-(-0.0924)(28.0)}{9} = 1.6096$

$y = 1.6096 - 0.0924x$

H_0: $\beta_1 = 0$; H_a: $\beta_1 \neq 0$;

TS: $T = B_1/S_{B_1}$; RR: $|T| \geq 2.3646$

$t = -0.0924/0.0349 = -2.6441 \leq -2.3646$

There is evidence to suggest that $\beta_1 \neq 0$, the regression line is significant.

c. Normal probability plot:

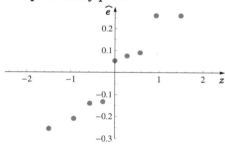

Scatter plot of residuals versus x:

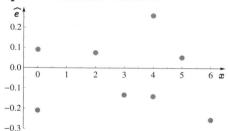

The points in the normal probability plot do not appear to lie along a straight line. This suggests a violation in the normality assumption. There is nothing unusual in the scatter plot of the residuals versus x.

12.119 a. Scatter plot (crack length, y versus cycles, x):

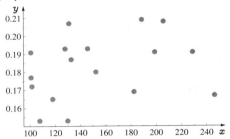

There does not appear to be a linear relationship. The scatter plot appears random.

b. $\widehat{\beta}_1 = \frac{(17)(478.6308)-(2600.3)(3.1060)}{(17)(4328.01)-(2600.3)^2} = 0.0001$

$\widehat{\beta}_0 = \frac{3.1060-(0.0001)(2600.3)}{17} = 0.1673$

$y = 0.1673 + 0.0001x$

ANOVA Summary Table:

Source of variation	Sum of squares	Degrees of freedom	Mean square	F	p value
Regression	0.0004	1	0.0004	1.15	0.3008
Error	0.0047	15	0.0003		
Total	0.0050	16			

c. The F test is not significant. There is no evidence to suggest a significant linear relationship. This supports the graph in part (a).

12.121 a. $y = 13.0865 + 0.0220x_1 - 0.0563x_2$

b. ANOVA Summary Table:

Source of variation	Sum of squares	Degrees of freedom	Mean square	F	p value
Regression	34.3151	2	17.1576	16.73	0.0001
Error	17.4304	17	1.0253		
Total	51.7455	19			

$H_0: \beta_1 = \beta_2 = 0$;
$H_a: \beta_i \neq 0$ for at least one i;
TS: $F = MSR/MSE$; RR: $F \geq 3.59$ ($\alpha = 0.05$)

$f = 17.1576/1.0253 = 16.73 \geq 3.59$

There is evidence to suggest that at least one of the regression coefficients is different from 0. The overall regression is significant.

$p = P(F \geq 16.73) = 0.0001$

$r^2 = 34.3151/51.7455 = 0.6632$

Approximately 66% of the variation in the data is explained by the regression model.

c. β_1: $t = 0.0220/0.0064 = 3.4520 \geq 2.4581$
$p < 0.0001$

β_2: $t = -0.0563/0.0150 = -3.7559 \leq -2.4581$
$p = 0.0016$

There is evidence to suggest both predictor variables are significant.

d. df $= 17$, $t_{0.005} = 2.8982$

$(\widehat{\beta}_0 + \widehat{\beta}_1 x^*) \pm t_{0.005}\sqrt{s^2 + s_{Y*}^2}$

$= 14.487 \pm (2.8982)\sqrt{1.0126^2 + 0.653^2}$

$= (10.994, 17.980)$

e. Solve for x_1:

$15 = 13.0865 + 0.0220x_1 - 0.0563(50)$

$\Rightarrow x_1 = 214.932$

12.123 a. Scatter plot (per-team payout, y versus total wins, x):

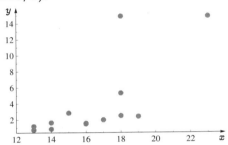

The relationship appears to be (positive) linear, with the exception of two outliers.

b. $\widehat{\beta}_1 = \frac{(14)(993.7240)-(227.0)(53.1790)}{(14)(3787.0)-(227.0)^2} = 1.2361$

$\widehat{\beta}_0 = \frac{53.1790-(1.2361)(27.0)}{14} = -16.2434$

$y = -16.2434 + 1.2361x$

ANOVA Summary Table:

Source of variation	Sum of squares	Degrees of freedom	Mean square	F	p value
Regression	162.50	1	162.50	13.67	0.0031
Error	142.68	12	11.89		
Total	305.18	13			

There is evidence to suggest the total number of wins can be used to predict the per-team payout. The overall test is significant, $p = 0.0031$.

c. df $= 12$, $t_{0.025} = 2.1788$

$(\widehat{\beta}_0 + \widehat{\beta}_1 x^*) \pm t_{0.025}s\sqrt{\frac{1}{n} + \frac{(x^*-\bar{x})^2}{S_{xx}}}$

$= 8.4779 \pm (2.1788)(3.4482)\sqrt{\frac{1}{14} + \frac{(20-16.2143)^2}{106.3571}}$

$= (5.0665, 11.8893)$

Section 12.6: The Polynomial and Qualitative Predictor Models

12.125

a. $Y_i = \beta_0 + \beta_1 x_{1i} + \beta_2 x_{2i} + \beta_3 x_{3i} + \beta_4 x_{1i} x_{2i}$
$\quad + \beta_5 x_{1i} x_{3i} + \beta_6 x_{2i} x_{3i} + E_i$

b. $Y_i = \beta_0 + \beta_1 x_{1i} + \beta_2 x_{1i}^2 + \beta_3 x_{1i}^3 + \beta_4 x_{1i}^4$
$\quad + \beta_5 x_{2i} + \beta_6 x_{3i} + \beta_7 x_{4i} + E_i$

c. $Y_i = \beta_0 + \beta_1 x_{1i} + \beta_2 x_{1i}^2 + \beta_3 x_{2i} + \beta_4 x_{3i} + E_i$

d. $Y_i = \beta_0 + \beta_1 x_{1i} + \beta_2 x_{1i}^2 + \beta_3 x_{1i}^3 + \beta_4 x_{2i}$
$\quad + \beta_5 x_{2i}^2 + \beta_6 x_{2i}^3 + \beta_7 x_{1i} x_{2i} + E_i$

12.127

a. $E(Y \mid 21) = 179.7 - 17.2(21) + 0.42(21)^2$
$\quad = 3.72$

b. Minimum value occurs at
$x = -(-17.2)/(2 * 0.42) = 20.4762$
$y = 179.7 - 17.2(20.4762) + 0.42(20.4762)^2 = 3.6048$

c. The values of y increase (along a parabola) from 12.2 to 41.7.

d. For $x_1 = 15$, $Y \sim N(16.2, 2.2^2)$
$P(Y > 18) = P(Z > 0.8182) = 0.2066$

12.129 a. Scatter plot (y versus x):

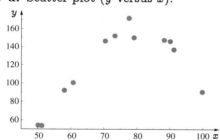

Model: $Y_i = \beta_0 + \beta_1 x_i + \beta_2 x_i^2 + E_i$

Estimated regression equation:
$y = -693.2870 + 21.5883x - 0.1370x^2$

b. $E(Y) = -693.2870 + 21.5883(65) - 0.1370(65)^2$
$\quad = 131.3187$

c.

x_i	y_i	$\widehat{y}_i$	$\widehat{e}_i$
99.8	90.4	97.1521	-6.7521
79.1	150.3	157.4482	-7.1482
90.1	146.2	140.0198	6.1802
58.0	92.0	98.1184	-6.1184
60.7	100.5	112.5143	-12.0143
77.6	171.9	157.2569	14.6431
51.2	53.8	53.0150	0.7850
50.1	54.2	44.5287	9.6713
88.3	147.6	145.1397	2.4603
70.4	146.5	147.7598	-1.2598
91.2	137.0	136.4541	0.5459
73.3	152.3	153.2929	-0.9929

Normal probability plot:

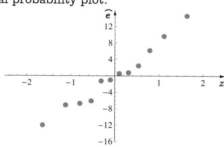

The points do not appear to fall along a straight line. There is some evidence to suggest the random error terms are not normal.

d. Scatter plot of residuals versus x:

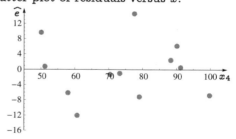

This plot suggests no there are no violations in the regression assumptions.

12.131 a. Estimated regression equation:
$y = -349.3143 + 124.5861x_1 - 5.9031x_2 - 6.2634x_1 x_2$
$H_0: \beta_1 = \beta_2 = \beta_3 = 0$;
$H_a: \beta_i \neq 0$ for at least one i;
TS: $F = \text{MSR}/\text{MSE}$; RR: $F \geq 3.24$
$f = 26440.49/2197.95 = 12.0296 \geq 3.24$

There is evidence to suggest at least one of the regression coefficients is different from 0.

β_1: $t = 124.5861/109.0954 = 1.1420$
$p = 0.2703$. There is no evidence to suggest $\beta_1 \neq 0$.
β_2: $t = -5.9031/25.0334 = -0.2358$

$p = 0.8166$. There is no evidence to suggest $\beta_2 \neq 0$.

β_3: $t = -6.2634/6.9953 = -0.8954$

$p = 0.3839$. There is no evidence to suggest $\beta_3 \neq 0$.

These tests suggest neither x_1 nor x_2 help to explain the variation in the dependent variable.

b. Estimated regression equation:

$y = -9.6013 + 27.6782x_1 - 27.8739x_2$

H_0: $\beta_1 = \beta_2 = 0$;

H_a: $\beta_i \neq 0$ for at least one i

TS: $F = MSR/MSE$; RR: $F \geq 3.59$

$f = 38780/2172.31 = 17.8518 \geq 3.59$

There is evidence to suggest at least one of the regression coefficients is different from 0.

β_1: $t = 27.6782/13.6291 = 2.0308$

$p = 0.0582$. The p value is very close to 0.05. There is some evidence to suggest $\beta_1 \neq 0$.

β_2: $t = -27.8739/4.9274 = -5.6570$

$p < 0.0001$. There is evidence to suggest $\beta_2 \neq 0$.

These tests suggest both x_1 and x_2 can be used to explain a significant amount of the variation in y.

c. $y^* \pm t_{0.025} \, s_{Y^*}$

$= -364.2 \pm (2.1098)(13.0) = (-391.6, -336.8)$

12.133 a. Scatter plot (y versus x):

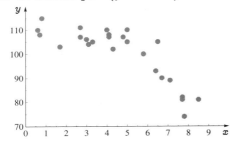

Model: $Y_i = \beta_0 + \beta_1 x_i + \beta_2 x_i^2 + E_i$

b. $y = 106.1325 + 3.5605x - 0.8364x^2$

c. H_0: $\beta_1 = \beta_2 = 0$;

H_a: $\beta_i \neq 0$ for at least one i;

TS: $F = MSR/MSE$; RR: $F \geq 3.44$

$f = 1276.13/22.45 = 56.8346 \geq 3.44$

There is evidence to suggest at least one of the regression coefficients is different from 0.

d. $E(Y) = 106.1325 + 3.5605(4) - 0.8364(4)^2$

$= 106.9911$

12.135 a. Scatter plot (y versus x):

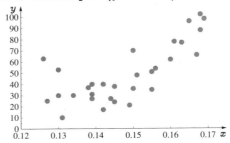

There appears to be a slight quadratic relationship between x and y.

b. Model: $Y_i = \beta_0 + \beta_1 x_i + \beta_2 x_i^2 + E_i$

Estimated regression equation:

$y = 1426.8290 - 20165.7205x + 72795.3363x^2$

c. $y^* \pm t_{0.025} s_{Y^*}$

$= 39.87 \pm (2.0518)(3.51) = (32.66, 47.07)$

d. Normal probability plot:

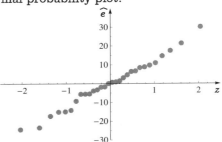

The points on the normal probability plot lie along a reasonably straight line. There is no evidence to suggest nonnormality.

12.137 a. Scatter plot (y versus x):

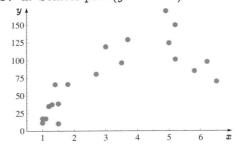

Model: $Y_i = \beta_0 + \beta_1 x_i + \beta_2 x_i^2 + E_i$

b. Estimated regression equation:

$y = -68.5940 + 91.4669x - 10.5621x^2$

c. H_0: $\beta_1 = \beta_2 = 0$;

H_a: $\beta_i \neq 0$ for at least one i;

TS: $F = MSR/MSE$; RR: $F \geq 6.11$

$f = 17944.7341/412.7579 = 43.48 \geq 6.11$

There is evidence to suggest at least one of the regression coefficients is different from 0.

d. Maximum value occurs at
$x = -91.4669/(2(-10.5621)) = 4.33$

12.139 a. ANOVA Summary Table:

Source of variation	Sum of squares	Degrees of freedom	Mean square	F	p value
Regression	1761.54	2	880.77	44.26	< 0.0001
Error	935.44	47	19.90		
Total	2696.98	49			

H_0: $\beta_1 = \beta_2 = 0$;
H_a: $\beta_i \neq 0$ for at least one i;
TS: $F = \text{MSR/MSE}$; RR: $F \geq 3.20$ ($\alpha = 0.05$)
$f = 880.77/19.90 = 44.26 \geq 3.20$
There is evidence to suggest at least one of the regression coefficients is different from 0.
$p = P(F \geq 44.26) = 0.0000000000155675$

b. β_2: $t = 2.22$, $p = 0.031 \leq 0.05$
There is evidence to suggest $\beta_2 \neq 0$. There is evidence to suggest gender does affect the relationship between depression and sports involvement.

c. $y = 34.634 - 2.074(4) + 2.81(0) = 26.338$

d. Female: Solve $34.634 - 2.074x \geq 20$ for x
$\Rightarrow x \geq 7.0559$
Male: Solve $34.634 - 2.074x + 2.81(1) \geq 20$ for x
$\Rightarrow x \geq 8.4108$

12.141 a. Model:
$Y_i = \beta_0 + \beta_1 x_{1i} + \beta_2 x_{2i} + \beta_3 x_{3i} + E_i$
where x_2 and x_3 are indicator variables with
N = (1, 0), W = (0, 1), and D = (0, 0).
Estimated regression equations:
$y = 0.0191 + 0.2309x_1 + 0.0009x_2 - 0.0093x_3$

b. H_0: $\beta_1 = \beta_2 = \beta_3 = 0$;
H_a: $\beta_i \neq 0$ for at least one i;
TS: $F = \text{MSR/MSE}$; RR: $F \geq 2.83$ ($\alpha = 0.05$)
$f = 0.0465/0.0034 = 13.57 \geq 2.83$
There is evidence to suggest at least one of the regression coefficients is different from 0. Therefore, there is evidence to suggest the overall regression is significant.

c. There is no evidence to suggest either β_2 or β_3 is different from 0.
β_2: $t = 0.0414$, $p = 0.9672$
β_3: $t = -0.4335$, $p = 0.6669$

d. Estimated regression equation:
$y = 0.0160 + 0.2312x_1$

This (reduced) simpler model is better. It includes the only significant independent variable and there is no significant decreases in the value of r^2.

12.143 a. Scatter plot (y versus x):

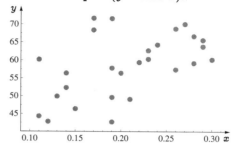

It appears as the proportion of fresh grass increases, the amount of milk produced per week also increases.

b. For H, $x_2 = 0$, and for G, $x_2 = 1$.
Estimated regression equation:
$y = 41.2246 + 77.5139x_1 + 2.0034x_2$

c. H_0: $\beta_1 = \beta_2 = 0$;
H_a: $\beta_i \neq 0$ for at least one i;
TS: $F = \text{MSR/MSE}$; RR: $F \geq 3.40$
$f = 291.7156/60.1257 = 4.8518 \geq 3.40$
There is evidence to suggest at least one of the regression coefficients is different from 0.
β_2: $t = 2.0034/3.0905 = 0.6482$
$p = 0.5230$
There is no evidence to suggest the coefficient on the indicator variable is different from 0.

d. Estimated regression equation:
$y = 42.5334 + 77.2836x_1$

This model is more appropriate since the predictor variable did not help to explain any significant additional variation in milk production.

12.145 a. ANOVA Summary Table:

Source of variation	Sum of squares	Degrees of freedom	Mean square	F	p value
Regression	1069.81	5	213.96	51.24	< 0.0001
Error	309.01	74	4.18		
Total	1378.82	79			

H_0: $\beta_1 = \beta_2 = \beta_3 = \beta_4 = \beta_5 = 0$;
H_a: $\beta_i \neq 0$ for at least one i;
TS: $F = \text{MSR/MSE}$; RR: $F \geq 3.28$
$f = 213.96/4.18 = 51.24 \geq 3.28$
There is evidence to suggest at least one of the regression coefficients is different from 0.

b. β_5: $t = 2.48$, $p = 0.015 \le 0.05$

There is evidence to suggest the coefficient on x_5 is different from 0. Therefore, there is evidence to suggest that gender does affect the weight of the child.

c.
$$E(Y) = 9.772 + 0.17033(88) - 1.9100(0) - 0.6444(0)$$
$$+ 0.9921(1) + 1.1749(1)$$
$$= 26.9280$$

d. The values of the indicator variables would change from $(0,0,0,0)$ to $(1,0,0,1)$. This would shift the regression line by $-1.9100 + 1.1749 = -0.7351$.

The values of the indicator variables would change from $(0,0,0,0)$ to $(0,0,1,1)$. This would shift the regression line by $0.9921 + 1.1749 = 2.167$

12.147 a. Scatter plot (number of people who had the flu, y, versus days since the first reported case, x):

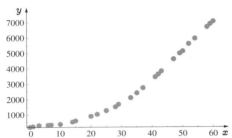

b. $\widehat{b} = \frac{(27)(429.2908) - (840.0)(39.1798)}{(27)(35902.0) - (848.0)^2} = -0.0808$

$\ln \widehat{a} = \frac{(39.1798)(-0.0808) - (840.0)}{27} = 3.9662$

$\widehat{a} = e^{3.9662} = 52.7836$

c. $y = \frac{10000}{1 + (52.7836)e^{(-0.0808)(80)}} = 9424.23$

d. Inflection point occurs at:

$x = \ln a / (-b) = 3.9662/0.0808 \approx 49$

5000 cases of the flu had been recorded at that time.

Estimated number of cases:

$y = \frac{10000}{1 + (52.7836)e^{(-0.0808)(49)}} = 4988$

Section 12.7: Model Selection Procedures

12.149 Suppose the variable x_i is eliminated from the model. Before eliminating another variable, consider a model with each of the variables already eliminated. Add (back to the model) the most significant variable, if one exists.

12.151 a. Scatter plot (r^2 versus k):

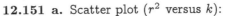
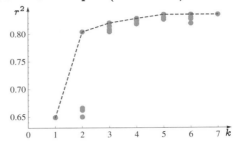

b. Scatter plot (s versus k):

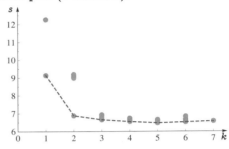

c. These graphs suggest the best predictor variables are $x1, x2, x3, x4, x7$. There is very little model improvement with any other subset.

12.153 a. Using forward selection, the model includes x_3 and x_4. Estimated regression equation:

$y = -24.0611 + 5.3227x_3 + 11.2561x_4$

b. The resulting model is the same. If x_1 or x_2 is added to the model, the regression coefficients on x_3 and x_4 are still significantly different from 0.

12.155 a. Using backward elimination, the estimated regression equation is:

$y = 9748.4975 + 0.0438x_1$

b. df $= 28$, $t_{0.025} = 2.0484$

$y^* \pm t_{0.025}s_{Y^*}$

$= 12041.3 \pm (2.0484)(82.4)$

$= (11872.5, 12210.0)$

12.157 a. Scatter plot (r^2 versus k):

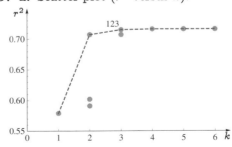

Include x_1, x_2, and x_3 in the model. There is little improvement in the model by adding additional variables.

b. Using backward elimination, include x_1 and x_2. Excluded first: x_6. Excluded second: x_5.

c. Using forward selection, include x_1 and x_2.

d. Considering all three methods, we should include x_1 and x_2 in the model. These two variables do make sense. The remaining variables all seem like they might effect hospital stay, especially x_3, the severity of the injury.

12.159 a. Backward elimination with $\alpha = 0.05$:
$y = 791.2332 + 2.3058x_4$

b. Forward selection with $\alpha = 0.05$:
$y = 1128.0157 + 0.2947x_5$

c. Different result with backward elimination ($\alpha = 0.10$):
$y = 777.5099 - 298.9328x_1 + 13.8035x_2 + 2.1006x_4$
Same result with forward selection ($\alpha = 0.10$):
$y = 1128.0157 + 0.2947x_5$

d. The value of r^2 for each of these models is very high. Therefore: use the simplest model via backward elimination, the model in part (a).

12.161 Use the indicator variables x_6 and x_7:
$x_2 = 0 = (0,0)$; $x_2 = 1 = (1,0)$; $x_2 = 2 = (0,1)$

a. Backward elimination, $\alpha = 0.10$:
$y = 4.5069 + 0.8415x_4$
Forward selection yields the same model.

b. Backward elimination, $\alpha = 0.10$:
$y = 7.8642 + 0.0001x_4^4$
Forward selection yields the same model.

c. $E(Y) = 7.8642 + 0.0001(18.5)^4 = 19.5777$

d. The resulting model includes only x_4^4. Consider a scatter plot of y versus x_4. This graph suggests a quartic relationship, with y increasing as x_4^4 increases. The variables x_4, x_4^2, and x_4^3 are not needed to describe the relationship.

12.163 a. Scatter plot (r^2 versus k):

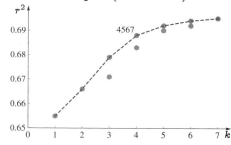

Include the variables x_4, x_5, x_6, and x_7. There is very little improvement in r^2 when an additional variable is added.

b. Using backward elimination, include the variables x_4, x_5, and x_6. Estimated regression equation:
$y = 103.27 - 2.01x_4 - 9.3x_5 + 0.76x_6$
As the United States propane stock increases, propane price decreases. As the New York spot heating oil price increases, propane price decreases. As the United States crude oil stock increases, propane price increases.

First variable excluded: x_2. Second variable excluded: x_1.

c. Using forward selection, include only x_4. Estimated regression equation:
$y = 171.5 - 2.04x_4$

d. Considering all three models, include the variables x_4, x_5, and x_6. This model is suggested by backward elimination, and there is very little increase in r^2 if an additional variable is added.

e. These procedures produce different results because it is likely that several predictor variables may be correlated.

Chapter Exercises, Optional Sections

12.165 a. Estimated regression equation:
$y = 108.1791 + 0.1708x_1 - 1.8284x_2$
As distance to the nearest fire hydrant increases, the insurance claim increases. As water pressure increases, the insurance claim decreases.

b. H_0: $\beta_1 = \beta_2 = 0$;
H_a: $\beta_i \neq 0$ for at least one i;
TS: $F = \text{MSR/MSE}$; RR: $F \geq 5.93$
$f = 24949.67/1181.38 = 21.12 \geq 5.93$
There is evidence to suggest at least one of the regression coefficients is different from 0.
$p = P(F \geq 21.12) = 0.00001484$

c. $r^2 = 49899.34/72345.61 = 0.6897$
Approximately 69% of the variation in y is explained by these two predictor variables.

d.
$E(Y) = 108.1791 + 0.1708(750) - 1.8284(50)$
$= 144.8916$

12.167 a. Forward selection, $\alpha = 0.05$, estimated regression equation:
$y = -154.3488 + 2.1751x_1 + 9.4304x_2 + 1.7348x_3 + 1.4421x_4$
As the diameter of the object, or the density of the

object, or the velocity of the object or the angle of the impact increases, the diameter of the crater created increases.

b.

$$E(Y) = -154.3488 + 2.1751(50) + 9.4304(8) \\ + 1.7348(60) + 1.4421(45) \\ = 198.8319$$

c. Normal probability plot:

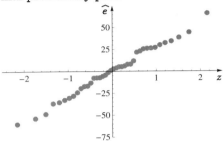

The points on the normal probability plot fall along a fairly straight line. There is no evidence to suggest nonnormality.

12.169 a. Estimated regression equation:
$$y = 60.0612 - 2.7552x_1 + 0.4536x_2 + 14.4641x_3$$

b. As the number of visits in the previous month increases, the amount of cash withdrawn decreases. As the amount of money in the user's account increases, the amount of cash withdrawn increases. If the day is a Friday, the amount of cash withdrawn increases. Pay days usually fall on a Friday. People often withdraw enough money on a Friday in preparation for the weekend.

c. ANOVA Summary Table:

Source of variation	Sum of squares	Degrees of freedom	Mean square	F	p value
Regression	8895.95	3	2965.32	20.76	< 0.0001
Error	5999.70	42	142.85		
Total	14895.65	45			

$p = P(F \geq 20.76) = 0.000000021$

d. β_1: $t = -2.7552/1.0430 = -2.6416$
$p = 0.0115$
There is evidence to suggest β_1 is different from 0.

β_2: $t = 0.4536/0.0993 = 4.5659$
$p < 0.0001$
There is evidence to suggest β_2 is different from 0.

β_3: $t = 14.4641/3.9589 = 3.6535$
$p = 0.0007$
There is evidence to suggest β_3 is different from 0.

e. df $= 42$, $t_{0.025} = 2.0181$

$$y^* \pm t_{0.025}\sqrt{s^2 + s_{Y*}^2} \\ = 72.58 \pm (2.0181)\sqrt{142.85 + 3.97^2} \\ = (47.16, 97.99)$$

f. Normal probability plot:

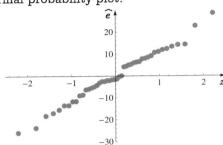

The points on this graph lie along a fairly straight line. There is no evidence of nonnormality.

12.171 a. Scatter plot (noise level, y versus resonant frequency, x):

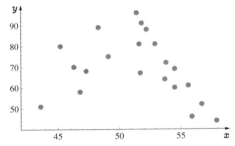

Model: $Y_i = \beta_0 + \beta_1 x_i + \beta_2 x_i^2 + E_i$

Estimated regression equation:
$$y = -1707.0161 + 71.4877x - 0.7142x^2$$

b. β_1: $t = 71.4877/13.8831 = 5.1493$
$p = 0.0001$
There is evidence to suggest β_1 is different from 0.

β_2: $t = -0.7142/0.1366 = -5.2282$
$p = 0.0001$
There is evidence to suggest β_2 is different from 0.

c. df $= 18$, $t_{0.025} = 2.1009$
$$y^* \pm t_{0.025}s_{Y*} \\ = 47.48 \pm (2.1009)(4.79) \\ = (37.42, 57.54)$$

d. Maximum frequency occurs at
$$x = -71.4877/(2(-0.7142)) = 50.05$$

e. Normal probability plot:

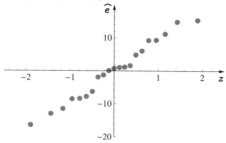

Scatter plot of residuals versus x:

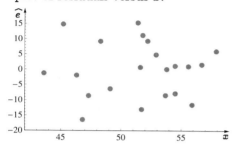

There is no evidence of a violation in the regression assumptions.

12.173 a. Forward selection, $\alpha = 0.05$:
$y = 22.9617 + 1.2238x_1 + 7.3375x_5$

b. The impedance should be high and the form factor should be headphone.

c. $E(Y) = 22.9617 + 1.2238(18) + 7.3375(1)$
$\qquad = 52.3276$

d. Normal probability plot:

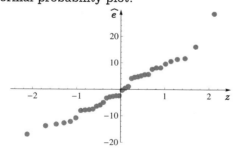

There is some evidence of nonnormality. The plot is not very linear and there appears to be an outlier.

e. Forward selection, $\alpha = 0.05$:
$y = 26.0049 + 0.9203x_1 + 9.7847x_5$

The resulting model includes the same two predictor variables. However, they were entered into the model in the opposite order and r^2 is greater with observation 21 removed.

12.175 a. Estimated regression equation:
$y = 6.8413 + 0.3744x_1 - 0.3413x_2 + 1.8672x_3$

b. H_0: $\beta_1 = \beta_2 = \beta_3 = 0$;
H_a: $\beta_i \neq 0$ for at least one i;
TS: $F = \text{MSR}/\text{MSE}$; RR: $F \geq 2.86$ ($\alpha = 0.05$)
$f = 75.49/1.98 = 38.04 \geq 2.86$

There is evidence to suggest at least one of the regression coefficients is different from 0.

β_1: $t = 0.3744/0.0943 = 3.9705$
$p = 0.0003 \leq 0.05$
There is evidence to suggest β_1 is different from 0.

β_2: $t = -0.3413/0.3987 = -0.8559$
$p = 0.3976$
There is no evidence to suggest β_2 is different from 0.

β_3: $t = 1.8672/0.7472 = 2.4990$
$p = 0.0170 \leq 0.05$
There is evidence to suggest β_3 is different from 0.

c. Estimated regression equation:
$y = 4.6814 + 0.3872x_1 + 1.8254x_3$

d. Dark clay:
$E(Y) = 4.6814 + 0.3872(3.5) + 1.8254(0) = 6.04$
light clay:
$E(Y) = 4.6814 + 0.3872(5.5) + 1.8254(1) = 8.64$

12.177 Use the indicator variables x_8 and x_9:
$x_3 = 0 = (0,0)$, $x_3 = 1 = (1,0)$, $x_3 = 2 = (0,1)$.

a. Estimated regression equation:
$y = 0.2922 + 0.0006x_1 + 0.0536x_2$
$\qquad - 0.0049x_4 - 0.0004x_5 + 0.0602x_6$
$\qquad +0.0003x_7 + 0.0077x_8 + 0.1051x_9$

The important variables as suggested by the t tests for significant regression coefficients: x_2, x_6.

b. Forward selection yields the same model as in part (a). Estimated regression equation:
$y = 0.3305 + 0.0531x_2 + 0.0566x_6$

c. Backward elimination also yields the same model as in part (a).

d. As street width increases, accidents per mile per year also increase. As the mean number of vehicles per day increases, accidents per mile per year increases.

e. Normal probability plot:

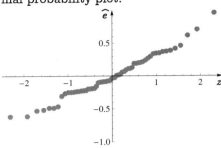

The points line along a fairly straight line. There is no overwhelming evidence to suggest the residuals are not normal.

Chapter 13: Categorical Data and Frequency Tables

Section 13.1: Univariate Categorical Data, Goodness-of-Fit Tests

13.1

Category	Expected cell count
1	$(218)(0.25) = 54.5$
2	$(218)(0.40) = 87.2$
3	$(218)(0.20) = 43.6$
4	$(218)(0.15) = 32.7$

13.3 a. H_0: $p_1 = 0.4$, $p_2 = 0.3$, $p_3 = 0.2$, $p_4 = 0.1$;
H_a: $p_i \neq p_{i0}$ for at least one i;
TS: $X^2 = \sum_{i=1}^{4} (n_i - e_i)^2/e_i$; RR: $X^2 \geq 11.3449$

b.

Category	Expected cell count
1	$(300)(0.4) = 120 \geq 5$
2	$(300)(0.3) = 90 \geq 5$
3	$(300)(0.2) = 60 \geq 5$
4	$(300)(0.1) = 30 \geq 5$

c. $\chi^2 = \frac{(115-120)^2}{120} + \cdots + \frac{(30-30)^2}{30} = 2.1528$
There is no evidence to suggest any one of the population proportions differs from its hypothesized value.

13.5 H_0: $p_i = 0.20$;
H_a: $p_i \neq p_{i0}$ for at least one i;
TS: $X^2 = \sum_{i=1}^{5} (n_i - e_i)^2/e_i$; RR: $X^2 \geq 9.4877$

$\chi^2 = \frac{(140-150)^2}{150} + \frac{(135-150)^2}{150} + \frac{(155-150)^2}{150}$
$+ \frac{(152-150)^2}{150} + \frac{(168-150)^2}{150}$
$= 4.52$

There is no evidence to suggest any one of the population proportions differs from its hypothesized value.
df $= 4$
$4.52 < 7.7794 \Rightarrow \chi^2 < \chi^2_{0.10} \Rightarrow p > 0.10$

13.7 H_0: $p_i = 0.20$;
H_a: $p_i \neq p_{i0}$ for at least one i;
TS: $X^2 = \sum_{i=1}^{5} (n_i - e_i)^2/e_i$; RR: $X^2 \geq 9.4877$

$\chi^2 = \frac{(185-200)^2}{200} + \frac{(180-200)^2}{200} + \frac{(230-200)^2}{200}$
$+ \frac{(220-200)^2}{200} + \frac{(185-200)^2}{200}$
$= 10.75 \geq 9.4877$

There is evidence to suggest at least one of the population proportions differs from its hypothesized value. There is evidence to suggest that customers who use flavor shots prefer one over the others.

13.9 H_0: $p_1 = 0.25$, $p_2 = 0.10$, $p_3 = 0.50$, $p_4 = 0.15$;
H_a: $p_i \neq p_{i0}$ for at least one i;
TS: $X^2 = \sum_{i=1}^{4} (n_i - e_i)^2/e_i$; RR: $X^2 \geq 7.8147$

$\chi^2 = \frac{(121-125)^2}{125} + \frac{(65-75)^2}{75} + \frac{(242-50)^2}{50} + \frac{(72-250)^2}{250}$
$= 5.0040$

There is no evidence to suggest any one of the population proportion eye colors differs from its hypothesized value.

13.11 H_0: $p_1 = 0.16$, $p_2 = 0.18$, $p_3 = 0.03$,
$p_4 = 0.07$, $p_5 = 0.21$, $p_6 = 0.35$;
H_a: $p_i \neq p_{i0}$ for at least one i;
TS: $X^2 = \sum_{i=1}^{6} (n_i - e_i)^2/e_i$; RR: $X^2 \geq 15.0863$

$\chi^2 = \frac{(90-69.76)^2}{69.76} + \frac{(82-78.48)^2}{78.48} + \frac{(21-13.08)^2}{13.08}$
$+ \frac{(33-30.52)^2}{30.52} + \frac{(80-91.56)^2}{91.56} + \frac{(130-152.60)^2}{152.60}$
$= 15.8340 \geq 15.0863$

There is evidence to suggest at least one of the population proportion act types differs from its hypothesized value.

13.13 H_0: $p_1 = 0.25$, $p_2 = 0.19$, $p_3 = 0.13$,
$p_4 = 0.07$, $p_5 = 0.09$, $p_6 = 0.23$;
H_a: $p_i \neq p_{i0}$ for at least one i;
TS: $X^2 = \sum_{i=1}^{6} (n_i - e_i)^2/e_i$; RR: $X^2 \geq 11.0705$

$\chi^2 = \frac{(120-126)^2}{126} + \frac{(95-95.76)^2}{95.76} + \frac{(70-65.52)^2}{65.52}$
$+ \frac{(58-55.44)^2}{55.44} + \frac{(60-45.36)^2}{45.36} + \frac{(101-115.92)^2}{115.92}$
$= 7.3617$

There is no evidence to suggest any one of this year's population proportions differs from last years.

13.15 H_0: $p_1 = 0.292$, $p_2 = 0.261$, $p_3 = 0.099$,
$p_4 = 0.077$, $p_5 = 0.271$;
H_a: $p_i \neq p_{i0}$ for at least one i;
TS: $X^2 = \sum_{i=1}^{5} (n_i - e_i)^2/e_i$; RR: $X^2 \geq 14.8603$

$\chi^2 = \frac{(125-127.896)^2}{127.896} + \frac{(96-114.318)^2}{114.318} + \frac{(45-43.362)^2}{43.362}$
$+ \frac{(44-33.726)^2}{33.726} + \frac{(128-118.698)^2}{118.698}$
$= 6.9214$

There is no evidence to suggest a shift in the proportion of applications by California location.

13.17 H_0: $p_1 = 0.14$, $p_2 = 0.29$, $p_3 = 0.44$, $p_4 = 0.13$;

H_a: $p_i \neq p_{i0}$ for at least one i;

TS: $X^2 = \sum\limits_{i=1}^{4} (n_i - e_i)^2/e_i$; RR: $X^2 \geq 11.3449$

$\chi^2 = \frac{(131-138.88)^2}{138.88} + \frac{(295-287.68)^2}{287.68} + \frac{(456-436.48)^2}{436.48}$
$+ \frac{(110-128.96)^2}{128.96}$

$= 4.2939$

There is no evidence to suggest any one of the literacy level population proportions has changed.

13.19 H_0: $p_i = 0.10$;

H_a: $p_i \neq p_{i0}$ for at least one i;

TS: $X^2 = \sum\limits_{i=1}^{10} (n_i - e_i)^2/e_i$; RR: $X^2 \geq 16.9190$

$\chi^2 = \frac{(38-50)^2}{50} + \frac{(45-50)^2}{50} + \frac{(55-50)^2}{50}$
$+ \frac{(53-50)^2}{50} + \frac{(46-50)^2}{50} + \frac{(62-50)^2}{50}$
$+ \frac{(35-50)^2}{50} + \frac{(45-50)^2}{50} + \frac{(56-50)^2}{50}$
$+ \frac{(65-50)^2}{50}$

$= 17.48 \geq 16.9190$

There is evidence to suggest at least one of the population proportions differs from its hypothesized value; evidence to suggest one airport mall is preferred over the rest.

13.21

Company	Total 2008 sales	Proportion of 2008 sales
General Motors	2,955,860	0.2240
Ford Motor Company	1,942,041	0.1472
DaimlerChrysler	1,697,458	0.1287
Toyota Motors	2,217,660	0.1681
American Honda	1,428,765	0.1083
Nissan N. America	951,446	0.0721
Hyundai Motor	401,742	0.0305
Volkswagen	310,888	0.0236
BMW	303,190	0.0230
Kia Motors	273,397	0.0207
Other	712,116	0.0540

H_0: $p_1 = 0.2240$, $p_2 = 0.1472$, $p_3 = 0.1287$, $p_4 = 0.1681$, $p_5 = 0.1083$, $p_6 = 0.0721$, $p_7 = 0.0305$, $p_8 = 0.0236$, $p_9 = 0.0230$, $p_{10} = 0207$, $p_{11} = 0.0540$;

H_a: $p_i \neq p_{i0}$ for at least one i;

TS: $X^2 = \sum\limits_{i=1}^{11} (n_i - e_i)^2/e_i$; RR: $X^2 \geq 23.2093$

$\chi^2 = \frac{(440-410.183)^2}{410.183} + \frac{(279-269.496)^2}{269.496} + \frac{(255-235.555)^2}{235.555}$
$+ \frac{(320-307.743)^2}{307.743} + \frac{(175-198.269)^2}{198.269} + \frac{(130-132.031)^2}{132.031}$
$+ \frac{(53-55.749)^2}{55.749} + \frac{(35-43.142)^2}{43.142} + \frac{(33-42.073)^2}{42.073}$
$+ \frac{(30-37.939)^2}{37.939} + \frac{(81-98.820)^2}{98.820}$

$= 15.8617$

There is no evidence to suggest the proportion of sales by company has changed in 2009.

Section 13.2: Bivariate Categorical Data, Tests for Homogeneity and Independence

13.23 a. df $= (2)(3) = 6$, $\chi^2_{0.05} = 12.5916$

b. df $= (1)(5) = 5$, $\chi^2_{0.01} = 15.0863$

c. df $= (3)(2) = 6$, $\chi^2_{0.025} = 14.4494$

d. df $= (4)(2) = 8$, $\chi^2_{0.001} = 26.1245$

13.25

	Category 1	2	3	4	
Population 1	18	14	18	15	65
2	25	21	16	12	74
3	32	33	26	28	119
	75	68	60	55	258

13.27

	Variable 2 1	2	3	4	
Variable 1 1	235 (235.05)	267 (259.76)	245 (269.43)	386 (368.75)	1133
2	241 (226.13)	264 (249.90)	280 (259.21)	305 (354.76)	1090
3	228 (237.75)	254 (262.74)	270 (272.53)	394 (372.98)	1146
4	219 (244.06)	235 (247.61)	263 (256.82)	363 (351.50)	1080
	923	1020	1058	1448	4449

H_0: Variable 1 and variable 2 are independent.

H_a: Variable 1 and variable 2 are dependent.

TS: $X^2 = \sum\limits_{i=1}^{4} \sum\limits_{j=1}^{4} \frac{(n_{ij} - e_{ij})^2}{e_{ij}}$; RR: $X^2 \geq 16.9190$

$\chi^2 = \frac{(235-235.05)^2}{235.05} + \cdots + \frac{(363-351.50)^2}{351.50} = 16.8226$

There is no evidence to suggest the two categorical variables are dependent.

13.29

		Product				
		Krimpets	Cupcakes	Kandy Kakes	Creamies	
Grocery store	Giant	90 (97.43)	80 (84.19)	95 (100.74)	92 (74.64)	357
	Shaw	81 (79.15)	66 (68.39)	87 (81.83)	56 (60.63)	290
	Weis	94 (88.42)	83 (76.41)	92 (91.43)	55 (67.74)	324
		265	229	274	203	971

H_0: The true product proportions are the same for all stores.

H_a: The true product proportions are not the same for all stores.

TS: $X^2 = \sum_{i=1}^{3}\sum_{j=1}^{4} \frac{(n_{ij}-e_{ij})^2}{e_{ij}}$; RR: $X^2 \geq 16.8119$

$\chi^2 = \frac{(90-97.43)^2}{97.43} + \cdots + \frac{(55-67.74)^2}{67.74} = 9.2674$

There is no evidence to suggest the true proportion of each favorite differs by grocery store.

13.31

		Funding for CBC				
		Decrease	Maintain	Increase	Unsure	
Region	Atlantic	17 (17.09)	58 (62.75)	36 (28.58)	9 (11.59	120
	Quebec	27 (42.14)	167 (154.78)	65 (70.49)	37 (28.59)	296
	Ontario	59 (50.55)	191 (185.63)	79 (84.54)	26 (34.29)	355
	West	68 (61.22)	212 (224.85	106 (102.40)	44 (41.53)	430
		171	628	286	116	1201

H_0: The true funding opinion proportions are the same for all regions.

H_a: The true funding opinion proportions are not the same for all regions.

TS: $X^2 = \sum_{i=1}^{4}\sum_{j=1}^{4} \frac{(n_{ij}-e_{ij})^2}{e_{ij}}$; RR: $X^2 \geq 23.5894$

$\chi^2 = \frac{(17-17.09)^2}{17.09} + \cdots + \frac{(44-41.53)^2}{41.53} = 17.8692$

There is no evidence to suggest funding opinion differs by region.

13.33

		Wine		
		Red	White	
Food	Red meat	86 (72.98)	46 (59.02)	132
	Fish or poultry	50 (63.02)	64 (50.98)	114
		136	110	246

H_0: Food and wine are independent.

H_a: Food and wine are dependent.

TS: $X^2 = \sum_{i=1}^{2}\sum_{j=1}^{2} \frac{(n_{ij}-e_{ij})^2}{e_{ij}}$; RR: $X^2 \geq 7.8794$

$\chi^2 = \frac{(86-72.98)^2}{72.98} + \cdot + \frac{(64-50.98)^2}{50.98} = 11.2179 \geq 7.8794$

There is evidence to suggest that food and wine are dependent. This suggests that diners are still following the traditional food-and-wine pairings.

13.35

		Diet				
		Meat	Fish, poultry	Veg	Med	
Cancer	Yes	56 (45.36)	23 (48.18)	21 (12.06)	20 (14.40)	120
	No	700 (710.64)	780 (754.82)	180 (188.94)	220 (225.60)	1880
		756	803	201	240	2000

H_0: Risk of colon cancer and diet are independent.

H_a: Risk of colon cancer and diet are dependent.

TS: $X^2 = \sum_{i=1}^{2}\sum_{j=1}^{4} \frac{(n_{ij}-e_{ij})^2}{e_{ij}}$; RR: $X^2 \geq 16.2662$

$\chi^2 = \frac{(56-45.36)^2}{45.36} + \cdots + \frac{(220-225.60)^2}{225.60}$

$= 26.022 \geq 16.2662$

There is evidence to suggest the risk of colon cancer and diet are dependent.

$df = (1)(3) = 3$

$26.022 > 21.1075 \Rightarrow \chi^2 > \chi^2_{0.0001} \Rightarrow p < 0.0001$

13.37

	Water chemistry	Violation Filtration system	Policy / manage	
Hotel / motel	369 (371.93)	326 (326.99)	165 (161.08)	860
Condominium / apartments	1425 (1419.37)	1207 (1247.88)	650 (614.74)	3282
School / university	34 (38.92)	45 (34.22)	11 (16.86)	90
Private club	195 (200.67)	185 (176.42)	84 (86.91)	464
Wading / children's	258 (240.89)	209 (211.78)	90 (104.33)	557
Water park	31 (32.87)	40 (28.90)	5 (14.24)	76
Medical / therapy	12 (12.11)	11 (10.65)	5 (5.24)	28
Municipal	29 (35.46)	45 (31.18)	8 (15.36)	82
Camp- ground	39 (39.79)	35 (34.98)	18 (17.23)	92
	2392	2103	1036	5531

(Pool type label along left axis)

H_0: Type of violation and type of pool are independent.

H_a: Type of violation and type of pool are dependent.

TS: $X^2 = \sum_{i=1}^{9} \sum_{j=1}^{3} \frac{(n_{ij}-e_{ij})^2}{e_{ij}}$; RR: $X^2 \geq 31.9999$

$\chi^2 = \frac{(369-371.93)^2}{371.93} + \cdots + \frac{(18-17.23)^2}{17.23}$

$= 34.7235 \geq 31.9999$

There is evidence to suggest the type of violation and the type of pool are dependent.

$df = (8)(2) = 16$

$34.2672 \leq 34.7235 \leq 39.2524$

$\chi^2_{0.005} \leq \quad \chi^2 \quad \leq \chi^2_{0.001}$

$0.001 \leq \quad p \quad \leq 0.005$

13.39

	Response Very familiar	Somewhat familiar	Not that familiar	Not at all familiar	
US	340 (149.48)	1020 (835.89)	553 (894.00)	213 (246.63)	2126
G Brit.	33 (76.50)	402 (427.77)	457 (457.51)	196 (126.22)	1088
France	31 (73.48)	261 (410.87)	669 (439.43)	84 (121.23)	1045
Italy	32 (74.04)	547 (414.01)	400 (442.80)	74 (122.16)	1053
Spain	60 (70.80)	423 (395.93)	403 (423.45)	121 (116.82)	1007
Germany	21 (72.70)	238 (406.54)	610 (434.81)	165 (119.95)	1034
	517	2891	3092	853	7353

(Country label along left axis)

H_0: Familiarity with the UN and country are independent.

H_a: Familiarity with the UN and country are dependent.

TS: $X^2 = \sum_{i=1}^{6} \sum_{j=1}^{4} \frac{(n_{ij}-e_{ij})^2}{e_{ij}}$; RR: $X^2 \geq 30.5779$

$\chi^2 = \frac{(340-149.48)^2}{149.48} + \cdots + \frac{(165-119.95)^2}{119.95}$

$= 981.9745 \geq 30.5779$

There is evidence to suggest that familiarity with the UN and the country are dependent.

13.41

	Floss 1	2	3	4	5	6	7	
1	35 (42.45)	37 (40.95)	48 (46.51)	52 (52.71)	55 (59.87)	80 (73.45)	101 (92.06)	408
2	36 (40.47)	38 (39.04)	42 (44.34)	47 (50.26)	54 (57.09)	75 (70.03)	97 (87.77)	389
3	38 (42.34)	43 (40.85)	46 (46.40)	51 (52.58)	58 (59.73)	77 (73.27)	94 (91.83)	407
4	33 (37.14)	51 (35.83)	42 (40.70)	42 (46.12)	51 (52.39)	52 (64.27)	86 (80.55)	357
5	76 (66.27)	79 (63.93)	85 (72.61)	87 (82.30)	93 (93.48)	102 (114.68)	115 (143.73)	637
6	81 (59.51)	41 (57.41)	46 (65.20)	78 (73.90)	107 (83.94)	103 (102.98)	116 (129.06)	572
7	98 (108.82)	94 (104.98)	126 (119.24)	136 (135.14)	142 (153.50)	198 (188.31)	252 (236.01)	1046
	397	383	435	493	560	687	861	3816

(Brush label along left axis)

H_0: Flossing frequency and brushing frequency are independent.

H_a: Flossing frequency and brushing frequency are dependent.

TS: $X^2 = \sum_{i=1}^{7} \sum_{j=1}^{7} \frac{(n_{ij}-e_{ij})^2}{e_{ij}}$; RR: $X^2 \geq 58.6192$

$\chi^2 = \frac{(35-42.45)^2}{42.45} + \cdots + \frac{(252-236.01)^2}{236.01}$

$= 62.4388 \geq 58.6192$

There is evidence to suggest flossing frequency and brushing frequency are dependent.

$p = P(X^2 \geq 62.4388) = 0.0041$

Chapter Exercises

13.43 H_0: $p_i = 0.20$.
H_a: $p_i \neq p_{i0}$ for at least one i.
TS: $X^2 = \sum\limits_{i=1}^{5} (n_i - e_i)^2/e_i$; RR: $X^2 \geq 9.4877$

$\chi^2 = \frac{(46-41.6)^2}{41.6} + \frac{(54-41.6)^2}{41.6} + \frac{(32-41.6)^2}{41.6}$
$\quad + \frac{(30-41.6)^2}{41.6} + \frac{(46-41.6)^2}{41.6}$
$\quad = 10.0769 \geq 9.4877$

There is evidence to suggest at least one of the population proportion of sales of dog breed differs from 0.20.

13.45 H_0: $p_1 = 0.30$, $p_2 = 0.40$, $p_3 = 0.10$, $p_4 = 0.40$, $p_5 = 0.10$.
H_a: $p_i \neq p_{i0}$ for at least one i.
TS: $X^2 = \sum\limits_{i=1}^{5} (n_i - e_i)^2/e_i$; RR: $X^2 \geq 9.4877$

$\chi^2 = \frac{(103-95.1)^2}{95.1} + \frac{(119-126.8)^2}{126.8} + \frac{(40-31.70)^2}{31.70}$
$\quad + \frac{(35-31.7)^2}{31.7} + \frac{(20-31.7)^2}{31.7}$
$\quad = 7.9711$

There is no evidence to suggest any of the true population proportions differs from its hypothesized value.

13.47

Grocery shopping frequency

		Monthly	Weekly	Daily	Other	
Country	Col	102 (101.41)	112 (159.41)	100 (56.34)	90 (86.83)	404
	CR	101 (97.65)	189 (153.49)	28 (54.25)	71 (83.61)	389
	EC	103 (106.94)	180 (168.09)	42 (59.41)	101 (91.56)	426
		306	481	170	262	1219

H_0: The true proportions of grocery shopping frequency is the same for all countries.
H_a: The true proportions of grocery shopping frequency is not the same for all countries.
TS: $X^2 = \sum\limits_{i=1}^{3}\sum\limits_{j=1}^{4} \frac{(n_{ij}-e_{ij})^2}{e_{ij}}$; RR: $X^2 \geq 18.5476$

$\chi^2 = \frac{(102-101.41)^2}{101.41} + \cdots + \frac{(101-91.56)^2}{91.56}$
$\quad = 78.0454 \geq 18.5476$

There is overwhelming evidence to suggest the true proportion of grocery shopping frequency is not the same for all countries.

13.49

Performance level

		Advanced	Proficient	Basic	Below basic	
District	BP	174 (197.95)	130 (117.98)	60 (51.07)	60 (57.01)	424
	SC	326 (302.05)	168 (180.02)	69 (77.93)	84 (86.99)	647
		500	298	129	144	1071

H_0: The true proportions of each performance level is the same for each school district.
H_a: The true proportions of each performance level is not the same for each school district.
TS: $X^2 = \sum\limits_{i=1}^{2}\sum\limits_{j=1}^{4} \frac{(n_{ij}-e_{ij})^2}{e_{ij}}$; RR: $X^2 \geq 11.3449$

$\chi^2 = \frac{(174-197.95)^2}{197.95} + \cdots + \frac{(84-86.99)^2}{86.99}$
$\quad = 9.6684$

There is no evidence to suggest the performance on the mathematics PSSA exam is associated with school district.

13.51

Survival status

		Died	Survived	
Class	First	122 (220.01)	203 (104.99)	325
	Second	167 (192.94)	118 (92.06)	285
	Third	528 (477.94)	178 (228.06)	706
	Crew	673 (599.11)	212 (285.89)	885
		1490	711	2201

H_0: Survival status and class are independent.
H_a: Survival status and class are dependent.
TS: $X^2 = \sum\limits_{i=1}^{4}\sum\limits_{j=1}^{2} \frac{(n_{ij}-e_{ij})^2}{e_{ij}}$; RR: $X^2 \geq 7.8147$

$\chi^2 = \frac{(122-220.01)^2}{220.01} + \cdots + \frac{(212-285.89)^2}{285.89}$
$\quad = 190.4011 \geq 7.8147$

There is overwhelming evidence to suggest an association between class and survival status.

13.53 a.

	Portfolio majority			
	Mutual funds	Stocks	Bonds	
Optimistic	28 (26.71)	62 (42.98)	42 (62.32)	132
Neutral	35 (31.36)	45 (50.47)	75 (73.17)	155
Pessimistic	24 (28.93)	33 (46.56)	86 (67.51)	143
	87	140	203	430

(Outlook labels the rows.)

H_0: Portfolio majority and outlook are independent.
H_a: Portfolio majority and outlook are dependent.

TS: $X^2 = \sum\limits_{i=1}^{3} \sum\limits_{j=1}^{3} \frac{(n_{ij}-e_{ij})^2}{e_{ij}}$; RR: $X^2 \geq 18.4668$

$\chi^2 = \frac{(28-26.71)^2}{26.71} + \cdots + \frac{(86-67.51)^2}{67.51}$

$\quad = 26.0201 \geq 18.4668$

There is evidence to suggest that portfolio majority and outlook for economic recovery are dependent.

b. df $= (2)(2) = 4$
$26.0201 > 23.5127 \Rightarrow \chi^2 > \chi^2_{0.0001} \Rightarrow p < 0.0001$

c. About half in stocks, 30% in bonds, and 20% in mutual funds.

Chapter 14: Nonparametric Statistics

Section 14.1: The Sign Test

14.1 a. X = the number of observations greater than 15. $x = 7$

b. X = the number of observations greater than 70. $x = 3$

c. X = the number of observations greater than 51.5. $x = 6$

d. X = the number of observations greater than 0. $x = 13$

14.3 TS: X = the number of differences greater than 0

$x = 16$, $p = P(X \geq 16) = 0.0022 \leq 0.05$

There is evidence to suggest $\tilde{\mu}_1 > \tilde{\mu}_2$.

14.5 H_0: $\tilde{\mu} = 1.38$; H_a: $\tilde{\mu} < 1.38$;
TS: X = the number of observations greater than 1.38

$x = 3$, $p = P(X \leq 3) = 0.0176 \leq 0.05$

There is evidence to suggest that the population median concentration of norephinephrine in women with PD is less than 1.38 nmol/L.

14.7 H_0: $\tilde{\mu} = 100$; H_a: $\tilde{\mu} > 100$;
TS: X = the number of observations greater than 100

$x = 17$, $p = P(X \geq 17) = 0.0320 \leq 0.05$

There is evidence to suggest the median mileage is greater than 100.

14.9 H_0: $\tilde{\mu} = 42$; H_a: $\tilde{\mu} > 42$;
TS: X = the number of observations greater than 42

$x = 23$, $p = P(X \geq 23) = 0.1279$

There is no evidence to suggest the median age of sports fans has increased.

14.11 H_0: $\tilde{\mu}_1 - \tilde{\mu}_1 = 0$; H_a: $\tilde{\mu}_1 - \tilde{\mu}_2 \neq 0$;
TS: X = the number of observations greater than 0

$x = 14$, $p = 2P(X \geq 14) = 0.1153$

There is no evidence to suggest the median chlorophyll amount in surface water is different in April and August.

14.13 H_0: $\tilde{\mu}_1 - \tilde{\mu}_1 = 0$; H_a: $\tilde{\mu}_1 - \tilde{\mu}_2 > 0$;
TS: X = the number of observations greater than 0

$x = 13$, $p = P(X \geq 13) = 0.0037$

There is evidence to suggest the median VOC concentration is smaller when the scrubber is installed.

Section 14.2: The Signed-Rank Test

14.15 a.

Observation	Difference	Absolute difference	Rank
41	−19	19	9.0
66	−7	7	4.0
36	6	6	3.0
72	−32	32	16.0
33	−24	24	11.5
22	17	17	8.0
24	12	12	5.0
36	−29	29	15.0
47	−27	27	14.0
53	−26	26	13.0
28	−38	38	18.0
77	−22	22	10.0
31	−36	36	17.0
34	−1	1	2.0
38	−24	24	11.5
59	0	0	1.0
60	−13	13	6.0
45	−15	15	7.0

b.

Observation	Difference	Absolute difference	Rank
21.4	1.4	1.4	16.0
19.8	−0.2	0.2	2.5
21.6	1.6	1.6	21.5
20.3	0.3	0.3	5.0
19.6	−0.4	0.4	8.0
19.2	−0.8	0.8	12.0
18.5	−1.5	1.5	18.5
19.4	−0.6	0.6	10.0
20.1	0.1	0.1	1.0
18.8	−1.2	1.2	14.0
20.8	0.8	0.8	12.0
18.5	−1.5	1.5	18.5
21.5	1.5	1.5	18.5
19.6	−0.4	0.4	8.0
18.5	−1.5	1.5	18.5
21.6	1.6	1.6	21.5
18.7	−1.3	1.3	15.0
19.7	−0.3	0.3	5.0
20.8	0.8	0.8	12.0
20.4	0.4	0.4	8.0
19.7	−0.3	0.3	5.0
21.9	1.9	1.9	23.0
20.2	0.2	0.2	2.5

c.

Observation	Difference	Absolute difference	Rank
−2	3.0	3.0	16.5
−9	−4.0	4.0	23.0
−4	1.0	1.0	5.0
−8	−3.0	3.0	16.5
−7	−2.0	2.0	9.5
−6	−1.0	1.0	5.0
−9	−4.0	4.0	23.0
−9	−4.0	4.0	23.0
−7	−2.0	2.0	9.5
−8	−3.0	3.0	16.5
−7	−2.0	2.0	9.5
−7	−2.0	2.0	9.5
−7	−2.0	2.0	9.5
−5	0.0	0.0	2.0
−2	3.0	3.0	16.5
−2	3.0	3.0	16.5
−5	0.0	0.0	2.0
−2	3.0	3.0	16.5
−8	−3.0	3.0	16.5
−1	4.0	4.0	23.0
−8	−3.0	3.0	16.5
−5	0.0	0.0	2.0
−3	2.0	2.0	9.5
−4	1.0	1.0	5.0
−1	4.0	4.0	23.0

d.

Observation	Difference	Absolute difference	Rank
296	−9.4	9.4	10.0
250	−55.4	55.4	30.0
279	−26.4	26.4	15.0
264	−41.4	41.4	23.5
303	−2.4	2.4	3.0
308	2.6	2.6	4.0
267	−38.4	38.4	22.0
278	−27.4	27.4	16.0
271	−34.4	34.4	20.0
315	9.6	9.6	11.0
312	6.6	6.6	7.5
255	−50.4	50.4	28.0
263	−42.4	42.4	25.0
264	−41.4	41.4	23.5
310	4.6	4.6	6.0
285	−20.4	20.4	14.0
288	−17.4	17.4	13.0
254	−51.4	51.4	29.0
309	3.6	3.6	5.0
272	−33.4	33.4	19.0
312	6.6	6.6	7.5
258	−47.4	47.4	27.0
273	−32.4	32.4	18.0
307	1.6	1.6	2.0
260	−45.4	45.4	26.0
274	−31.4	31.4	17.0
269	−36.4	36.4	21.0
305	−0.4	0.4	1.0
314	8.6	8.6	9.0
293	−12.4	12.4	12.0

14.17 $H_0: \tilde{\mu}_1 - \tilde{\mu}_2 = 0$; $H_a: \tilde{\mu}_1 - \tilde{\mu}_2 \neq 0$;
TS: $T_+ =$ the sum of the ranks corresponding to the positive differences $d_i - 0$;
RR: $T_+ \leq 43$ or $T_+ \geq 167$
$t_+ = 150.5$, $p = 2P(T_+ \geq 150.5) = 0.0896$
There is no evidence to suggest $\tilde{\mu}_1$ is different from $\tilde{\mu}_2$.

14.19 $H_0: \tilde{\mu} = 3$; $H_a: \tilde{\mu} \neq 3$;
TS: $T_+ =$ the sum of the ranks corresponding to the positive differences $x_i - 3$;
RR: $T_+ \leq 37$ or $T_+ \geq 173$

$t_+ = 30.5 \leq 37$
$p = 2P(T_+ \leq 30.5) = 0.0036 \leq 0.01$

There is evidence to suggest the median renal blood-flow rate is different from 3.

14.21 $H_0: \tilde{\mu}_1 - \tilde{\mu}_2 = 0$; $H_a: \tilde{\mu}_1 - \tilde{\mu}_2 > 0$;
TS: $T_+ =$ the sum of the ranks corresponding to the positive differences $d_i - 0$;
RR: $T_+ \geq 131$

$t_+ = 149.5 \geq 131$

There is evidence to suggest the intervention program decreased the median A1C value.

14.23 $H_0: \tilde{\mu}_1 - \tilde{\mu}_2 = 0$; $H_a: \tilde{\mu}_1 - \tilde{\mu}_2 \neq 0$;

TS: $Z = \frac{T_+ - \mu_{T_+}}{\sigma_{T_+}}$; RR: $|Z| \geq z_{0.05} = 1.6449$

$z = \frac{214.5 - 264.0}{\sqrt{2860.0}} = -0.9256$

There is no evidence to suggest a difference in median collection times.

14.25 $H_0: \tilde{\mu}_1 - \tilde{\mu}_2 = 0$; $H_a: \tilde{\mu}_1 - \tilde{\mu}_2 \neq 0$;
TS: $T_+ = $ the sum of the ranks corresponding to the positive differences $d_i - 0$;
RR: $T_+ \leq 29$ or $T_+ \geq 107$ ($n = 16$ for the test)

$t_+ = 102$

There is no evidence to suggest a difference in median on-time percentages. This suggests the on-time percentages have not changed from 2006 to 2007.

Section 14.3: The Rank-Sum Test

14.26 a.

Sample 1		Sample 2	
Obs	Rank	Obs	Rank
37	6	45	10
21	1	42	9
46	11	22	2
29	4	41	8
34	5	24	3
		39	7

b.

Sample 1		Sample 2	
Obs	Rank	Obs	Rank
4.5	15.0	4.6	16.0
1.8	5.0	6.6	18.0
3.4	13.0	1.2	2.0
1.4	3.0	6.0	17.0
2.2	7.5	2.4	10.0
2.1	6.0	2.3	9.0
1.5	4.0	2.2	7.5
3.7	14.0	0.4	1.0
		2.5	11.0
		2.7	12.0

c.

Sample 1		Sample 2	
Obs	Rank	Obs	Rank
820	11.0	850	25.0
809	4.0	840	21.0
872	33.0	813	6.0
826	15.0	842	23.0
814	7.0	870	30.0
887	39.0	839	20.0
825	14.0	888	40.0
884	38.0	816	8.0
876	35.0	822	13.0
862	27.0	879	36.0
858	26.0	821	12.0
846	24.0	865	28.5
841	22.0	832	19.0
801	2.0	865	28.5
892	41.0	818	9.5
871	31.5	827	16.0
882	37.0	899	42.0
803	3.0	818	9.5
		831	18.0
		830	17.0
		810	5.0
		875	34.0
		871	31.5
		800	1.0

14.29 a. $W \leq 7$, $\alpha = 0.0357$

b. $W \geq 24$, $\alpha = 0.0571$

c. $W \leq 16$ or $W \geq 44$, $\alpha = 0.0540$

d. $W \leq 27$, $\alpha = 0.0100$

e. $W \leq 44$ or $W \geq 75$, $\alpha = 0.1142$

f. $W \geq 90$, $\alpha = 0.0103$

14.31 $H_0: \tilde{\mu}_1 - \tilde{\mu}_2 = 0$; $H_a: \tilde{\mu}_1 - \tilde{\mu}_2 > 0$;
TS: $W = $ the sum of the ranks corresponding to the first (smaller) sample;
RR: $W \geq 63$

Sample 1		Sample 2	
Obs	Rank	Obs	Rank
51.8	1	57.2	7
55.6	3	61.1	13
57.6	8	58.2	9
58.7	10	56.9	5
56.2	4	57.1	6
53.1	2	63.3	14
		59.7	11
		60.6	12

$w = 1 + 3 + 8 + 10 + 4 + 2 = 28$

There is no evidence to suggest $\tilde{\mu}_1 > \tilde{\mu}_2$.

14.33 $H_0: \tilde{\mu}_1 - \tilde{\mu}_2 = 0$; $H_a: \tilde{\mu}_1 - \tilde{\mu}_2 \neq 0$;

TS: $W =$ the sum of the ranks corresponding to the first (smaller) sample;

RR: $W \leq 51$ or $W \geq 93$

Sample 1		Sample 2	
Obs	Rank	Obs	Rank
12.5	4.5	12.5	4.5
12.9	8.5	13.3	11.0
12.4	3.0	13.8	15.0
11.5	2.0	13.2	10.0
12.7	6.0	13.5	12.5
12.8	7.0	13.9	16.0
14.2	17.0	13.6	14.0
11.0	1.0	13.5	12.5
		12.9	8.5

$w = 4.5 + 8.5 + \cdots + 1.0 = 49 \leq 51$

$p = 2P(W \leq 49) = 0.0274 \leq 0.05$

There is evidence to suggest the population medians are different.

14.35 a. H_0: $\tilde{\mu}_1 - \tilde{\mu}_2 = 0$; H_a: $\tilde{\mu}_1 - \tilde{\mu}_2 < 0$;

TS: $W =$ the sum of the ranks corresponding to the first (smaller) sample;

RR: $W \leq 35$

Sample 1		Sample 2	
Obs	Rank	Obs	Rank
95	5.0	119	11.0
94	4.0	130	13.5
102	9.0	99	6.0
100	7.0	126	12.0
93	2.0	114	10.0
93	2.0	130	13.5
93	2.0	101	8.0

$w = 5.0 + 4.0 + \cdots + 2.0 = 31 \leq 35$

There is evidence to suggest the median impact strength is higher for the new jackhammer.

b. $p = P(W \leq 31) = 0.0030$

14.37 H_0: $\tilde{\mu}_1 - \tilde{\mu}_2 = 0$; H_a: $\tilde{\mu}_1 - \tilde{\mu}_2 \neq 0$;

TS: $Z = \frac{W - \mu_W}{\sigma_W}$; RR: $|Z| \geq z_{0.025} = 1.96$

Sample 1		Sample 2	
Obs	Rank	Obs	Rank
41	32.5	35	9.5
40	30.0	37	18.5
38	22.5	36	15.0
35	9.5	35	9.5
37	18.5	35	9.5
34	3.5	34	3.5
34	3.5	34	3.5
36	15.0	38	22.5
34	3.5	36	15.0
38	22.5	39	27.0
39	27.0	38	22.5
41	32.5	38	22.5
39	27.0	35	9.5
40	30.0	36	15.0
40	30.0	35	9.5
		34	3.5
		36	15.0
		38	22.5

$\mu_W = \frac{15(15+18+1)}{2} = 255.0$

$\sigma_W^2 = \frac{(15)(18)(15+18+1)}{12} = 765.0$

$w = 32.5 + 30.0 + \cdots + 30.0 = 307.5$

$z = \frac{307.5 - 255.0}{\sqrt{765.0}} = 1.8981$

There is no evidence to suggest the population median holding temperatures are different.

14.39 a. H_0: $\tilde{\mu}_1 - \tilde{\mu}_1 = 0$; H_a: $\tilde{\mu}_1 - \tilde{\mu}_2 > 0$;

TS: $X =$ the number of observations greater than 0;

RR: $X \geq 11$ ($\alpha = 0.0592$)

$x = 15 \geq 11$

There is overwhelming evidence to suggest the median amount of particulates after the smoking regulations is less than the median amount before the regulations.

b. H_0: $\tilde{\mu}_1 - \tilde{\mu}_2 = 0$; H_a: $\tilde{\mu}_1 - \tilde{\mu}_2 > 0$;

TS: $T_+ =$ the sum of the ranks corresponding to the positive differences $d_i - 0$;

RR: $T_+ \geq 89$ ($\alpha = 0.0535$)

$t_+ = 120 \geq 89$

There is evidence to suggest the median amount of particulates after the smoking regulations is less than the median amount before the regulations.

c. H_0: $\tilde{\mu}_1 - \tilde{\mu}_2 = 0$; H_a: $\tilde{\mu}_1 - \tilde{\mu}_2 > 0$;

TS: $Z = \frac{W - \mu_W}{\sigma_W}$; RR: $Z \geq 1.6449$

$\mu_W = \frac{15(15+15+1)}{2} = 232.5$

$\sigma_W^2 = \frac{(15)(15)(15+15+1)}{12} = 581.25$

$w = 27.0 + 28.0 + \cdots + 26.0 = 339.5$

$z = \frac{339.5 - 232.5}{\sqrt{581.25}} = 4.4382 \geq 1.6449$

There is evidence to suggest the median amount of particulates after the smoking regulations is less than the median amount before the regulations.

d. All three tests lead to the same conclusion. The rank-sum test, however, should not be used since the samples are dependent.

Section 14.4: The Kruskal-Wallis Test

14.41 H_0: The 5 samples are from identical populations.

H_a: At least two of the populations are different.

TS: $H = \left[\frac{12}{n(n+1)} \sum \frac{R_i^2}{n_i}\right] - 3(n+1)$

RR: $H \geq \chi_{0.01}^2 = 13.2767$

$h = \left[\frac{12}{(74)(75)} \left(\frac{395.5^2}{12} + \cdots + \frac{1044.0^2}{20}\right)\right]$
$\qquad - 3(75)$
$\qquad = 16.1591 \geq 13.2767$

There is evidence to suggest at least two of the populations are different.

14.43 H_0: The 3 samples are from identical populations.

H_a: At least two of the populations are different.

TS: $H = \left[\frac{12}{n(n+1)} \sum \frac{R_i^2}{n_i}\right] - 3(n+1)$

RR: $H \geq \chi_{0.05}^2 = 5.9915$

$h = \left[\frac{12}{(58)(59)} \left(\frac{456.5^2}{20} + \frac{729.0^2}{20} + \frac{525.5^2}{18}\right)\right]$
$\qquad - 3(59)$
$\qquad = 6.5184$

There is evidence to suggest at least two of the populations are different.

14.45 H_0: The 4 samples are from identical populations.

H_a: At least two of the populations are different.

TS: $H = \left[\frac{12}{n(n+1)} \sum \frac{R_i^2}{n_i}\right] - 3(n+1)$

RR: $H \geq \chi_{0.05}^2 = 7.8147$

$h = \left[\frac{12}{(57)(58)} \left(\frac{525.0^2}{14} + \frac{355.5^2}{15} + \frac{373.0^2}{12} + \frac{399.5^2}{16}\right)\right]$
$\qquad - 3(58)$
$\qquad = 6.3338$

There is no evidence to suggest the populations are different.

14.47 H_0: The 4 samples are from identical populations.

H_a: At least two of the populations are different.

TS: $H = \left[\frac{12}{n(n+1)} \sum \frac{R_i^2}{n_i}\right] - 3(n+1)$

RR: $H \geq \chi_{0.01}^2 = 11.3449$

$h = \left[\frac{12}{(32)(33)} \left(\frac{101.5^2}{8} + \frac{147.0^2}{8} + \frac{165.5}{8} + \frac{114.0^2}{8}\right)\right]$
$\qquad - 3(33)$
$\qquad = 3.6953$

There is no evidence to suggest the populations are different.

14.49 H_0: The 3 samples are from identical populations.

H_a: At least two of the populations are different.

TS: $H = \left[\frac{12}{n(n+1)} \sum \frac{R_i^2}{n_i}\right] - 3(n+1)$

RR: $H \geq \chi_{0.01}^2 = 5.9915$

$h = \left[\frac{12}{(18)(19)} \left(\frac{55.5^2}{6} + \frac{47.0^2}{6} + \frac{68.5^2}{6}\right)\right] - 3(19)$
$\qquad = 1.3713$

There is no evidence to suggest the uncompressed depth populations are different.

$df = 3 - 1 = 2$

$1.3713 < 4.6052 \Rightarrow h < \chi_{0.10}^2 \Rightarrow p > 0.10$

Section 14.5: The Runs Test

14.51 a. There are 8 runs in this sequence of observations.

b. There are 8 runs in this sequence of observations.

c. There are 11 runs in this sequence of observations.

d. There are 15 runs in this sequence of observations.

14.53 a. There are 8 runs in this sequence of observations.

b. There are 5 runs in this sequence of observations.

c. There are 8 runs in this sequence of observations.

d. There is 1 run in this sequence of observations.

14.55 H_0: The sequence of observations is random.

H_a: The sequence of observations is not random.

TS: $V =$ the number of runs;

RR: $V \leq 7$ or $V \geq 17$

$v = 13$

There is no evidence to suggest the order of observations is not random.

14.57 H_0: The sequence of observations is random.

H_a: The sequence of observations is not random.

TS: $V =$ the number of runs;

RR: $V \leq 3$ or $V \geq 8$ $(\alpha = 0.0385)$

$v = 5$

There is no evidence to suggest the order of observations is not random with respect to gender.

14.59 H_0: The sequence of observations is random.
H_a: The sequence of observations is not random.
TS: $Z = \frac{V - \mu_V}{\sigma_V}$; RR: $|Z| \geq 2.3263$

$\mu_V = \frac{2(16)(19)}{16+19} + 1 = 18.3714$

$\sigma_V^2 = \frac{2(16)(19)(2(16)(19)-16-19)}{(16+19)^2(16+19-1)} = 8.3646$

$z = \frac{22-18.3714}{\sqrt{8.3646}} = 1.2546$

There is no evidence to suggest the order of observations is not random.

14.61 H_0: The sequence of observations is random.
H_a: The sequence of observations is not random.
TS: $V =$ the number of runs;
RR: $V \leq 3$ or $V \geq 10$ $(\alpha = 0.0242)$

$v = 9$

There is no evidence to suggest the order of observations is not random.

14.63 Answers will vary. If the random number generator is good, then the number of times the null hypothesis is rejected should be close to 5.

Section 14.6: Spearman's Rank Correlation Coefficient

14.65 a. $r_S = 0.5357$. This indicates a moderate positive relationship.

b. $r_S = 0.3333$. This indicates a weak positive relationship.

c. $r_s = 0.0637$. This indicates there is no definitive relationship.

d. $r_S = -0.2922$. This indicates a weak negative relationship.

14.67 $r_S = 1 - \frac{6(8)}{6(36-1)} = 0.7714$

There is a positive relationship between x and y. As the price of a stateroom increases, so does the number of days before sailing. Therefore, this suggests that cruise prices are reduced at the last minute, although we cannot conclude this is a cause and effect relationship.

14.69 $r_S = 1 - \frac{6(808)}{15(225-1)} = -0.4429$

This value suggests there is a weak to moderate negative linear relationship between density and soil texture. As bulk density increases, the soil texture tends to decrease.

14.71 Find the sample correlation coefficient between the ranks since there are tied observations.

$r = \frac{-60.75}{\sqrt{(278.5)(272.0)}} = -0.2207$

This value suggests there is a weak negative linear relationship between price and number of weeks on the list. As the price increases, the number of weeks on the list tends to decrease.

14.73 a.

Sample 1		Sample 2		
Obs	Rank	Obs	Rank	d_i
71	9.0	64	7.5	1.5
69	6.5	56	1.0	5.5
66	3.0	65	9.5	-6.5
69	6.5	62	4.0	2.5
62	1.0	65	9.5	-8.5
67	4.0	64	7.5	-3.5
69	6.5	57	2.5	4.0
64	2.0	63	5.5	-3.5
73	10.0	57	2.5	7.5
69	6.5	63	5.5	1.0

b. $r = \frac{-46.5}{\sqrt{(77.5)(80.5)}} = -0.5887$

c. $r_S = 1 - \frac{6(251)}{10(100-1)} = -0.5212$

d. These values are different because there are tied observations. There is a moderate negative relationship. As shear strength increases, temperature cycle stress decreases.

Chapter Exercises

14.75 H_0: $\tilde{\mu} = 5$; H_a: $\tilde{\mu} > 5$;
TS: $X =$ the number of observations greater than 5

$x = 10$

$p = P(X \geq 10) = 0.1509$

There is no evidence to suggest the median nitrogen emissions amount is greater than 5.

14.77 H_0: $\tilde{\mu} = 2.00$; H_a: $\tilde{\mu} \neq 2.00$;
TS: $X =$ the number of observations greater than 2.0; RR: $X \leq 5$ or $X \geq 15$ $(\alpha = 0.0414)$

$x = 15 \geq 15$

$p = 2P(X \geq 15) = 0.0414 \leq 0.05$

There is evidence to suggest the median coverage amount is different from 2.

14.79 a. H_0: $\tilde{\mu} = 630$; H_a: $\tilde{\mu} < 630$;
TS: $T_+ =$ the sum of the ranks corresponding to the positive differences $x_i - 630$;
RR: $T_+ \leq 20$ $(\alpha = 0.0108)$

$t_+ = 17.5 \leq 20$

There is evidence to suggest the median spray height is less than 630.

b. $p = P(T_+ \leq 17.5) = 0.0062$

c. The distribution is not assumed to be symmetric.

14.81 H_0: $\tilde{\mu}_1 - \tilde{\mu}_2 = 0$; H_a: $\tilde{\mu}_1 - \tilde{\mu}_2 > 0$;
TS: $T_+ = $ the sum of the ranks corresponding to the positive differences $d_i - 0$;
RR: $T_+ \geq 100$ ($\alpha = 0.0523$)

$t_+ = 133 \geq 100$

There is evidence to suggest the median time spent working per week for those well off is greater than for those who just manage.

14.83 a. H_0: $\tilde{\mu}_1 - \tilde{\mu}_2 = 0$; H_a: $\tilde{\mu}_1 - \tilde{\mu}_2 \neq 0$;
TS: $W = $ the sum of the ranks corresponding to the first sample;
$W \leq 52$ or $W \geq 84$ ($\alpha = 0.1048$)

$w = 78.5$

There is no evidence to suggest there is a difference in the absorbed radiation by machine.

b. $p = 2P(W \geq 78.5) = 0.2786$

14.85 H_0: $\tilde{\mu}_1 - \tilde{\mu}_2 = 0$; H_a: $\tilde{\mu}_1 - \tilde{\mu}_2 \neq 0$;
TS: $Z = \frac{W - \mu_W}{\sigma_W}$; RR: $|Z| \geq 3.2905$

$\mu_W = \frac{20(20+22+1)}{2} = 430.0$

$\sigma_W^2 = \frac{(20)(22)(20+22+1)}{12} = 1576.6667$

$z = \frac{533 - 430}{\sqrt{1576.6667}} = 2.5940$

There is no evidence to suggest the land value for farmland is different in these two counties.

14.87 H_0: The 4 samples are from identical populations.
H_a: At least two of the populations are different.
TS: $H = \left[\frac{12}{n(n+1)} \sum \frac{R_i^2}{n_i} \right] - 3(n+1)$

RR: $H \geq \chi_{0.025}^2 = 9.3484$

$h = \left[\frac{12}{(69)(70)} \left(\frac{611.0^2}{15} + \frac{605.5^2}{16} + \frac{609.0^2}{18} + + \frac{589.5^2}{20} \right) \right]$
$\quad - 3(70)$

$\quad = 3.1241$

There is no evidence to suggest the transmitter power populations are different.

df $= 4 - 1 = 3$

$p = P(X^2 \geq 3.1241) = 0.3729$

14.89 a. H_0: The sequence of observations is random.
H_a: The sequence of observations is not random.
TS: $V = $ the number of runs;

RR: $V \leq 3$ or $V \geq 10$ ($\alpha = 0.0476$)

$v = 10 \geq 10$

There is evidence to suggest the order of observations is not random with respect to email password.

b. $p = 2P(V \geq 10) = 0.0476$.

14.91 H_0: The sequence of observations is random.
H_a: The sequence of observations is not random.
TS: $Z = \frac{V - \mu_V}{\sigma_V}$; RR: $|Z| \geq 2.5758$

$\mu_V = \frac{2(17)(23)}{17+23} + 1 = 20.55$

$\sigma_V^2 = \frac{2(17)(23)(2(17)(23)-17-23)}{(17+23)^2(17+23-1)} = 9.2988$

$z = \frac{26 - 20.55}{\sqrt{9.2988}} = 1.7872$

There is no evidence to suggest the order of observations is not random.

$p = 2P(Z \geq 1.7872) = 0.0739$

14.93 $r_S = 1 - \frac{6(404)}{12(144-1)} = -0.4126$

This value suggests there is a weak to moderate negative linear relationship between the quality score and the total number of people in the hospital. This suggests as the quality score increases, the total number of people in the hospital decreases.

14.95 a. H_0: $\tilde{\mu}_1 - \tilde{\mu}_2 = 0$; H_a: $\tilde{\mu}_1 - \tilde{\mu}_2 < 0$;
TS: The number of pairwise differences greater than 0; RR: $X \leq 5$ ($\alpha = 0.0207$)

$x = 20$

There is no evidence to suggest the median freon weight before service is less than the median freon weight after service.

b. Since we cannot reject the null hypothesis, there is no evidence to suggest that recharging an air conditioner increases the amount of freon in the system.

14.97 a. H_0: The sequence of observations is random.
H_a: The sequence of observations is not random.
TS: $V = $ the number of runs;
RR: $V \leq 4$ or $V \geq 11$ ($\alpha = 0.0709$)

$v = 7$

There is no evidence to suggest the order of observations is not random with respect to exterior finish.

b. Using the runs test, we cannot tell if the advertising campaign was successful. And, we don't know the historical proportion of home-builders who use vinyl. Therefore, we cannot tell if this proportion has increased.

14.99 a. H_0: The 3 samples are from identical populations.

H_a: At least two of the populations are different.

TS: $H = \left[\frac{12}{n(n+1)} \sum \frac{R_i^2}{n_i} \right] - 3(n + 1)$

RR: $H \geq \chi^2_{0.01} = 9.2103$

$h = \left[\frac{12}{(48)(49)} \left(\frac{572.5^2}{16} + \frac{466.5^2}{16} + \frac{137.0^2}{16} \right) \right] - 3(49)$

$\quad = 32.8940 \geq 9.2103$

There is excellent evidence to suggest at least two of the populations are different.

b. df $= 3 - 1 = 2$

$32.8940 > 18.4207 \Rightarrow h > \chi^2_{0.0001} \Rightarrow p < 0.0001$

c. The safest time to drive is *other times*.